Geology for Kids

*A Biblical Approach to
Earth Science and Earth History*

The Northwest Treasures Curriculum Project
Building Faith for a Lifetime of Faith

By Patrick Nurre

Geology for Kids
*A Biblical Approach to
Earth Science and Earth History*

By Patrick Nurre

Geology for Kids – A Biblical Approach to Earth Science and Earth History
Published by Northwest Treasures
Bothell, Washington
425-488-6848
NorthwestRockAndFossil.com
northwestexpedition@msn.com Copyright 2015 by
Patrick Nurre.
All rights reserved.
Printed in the United States of America. No part of this book may be reproduced in any manner whatsoever without written permission except in the case of brief quotations embodied in critical articles and reviews. Title page photo: by Vicki Nurre.
Scripture quotations taken from the New American Standard Bible®.
Copyright © 1960, 1962, 1963, 1968, 1971, 1972, 1973,
1975, 1977, 1995 by The Lockman Foundation
Used by permission. (www.Lockman.org)

Geology for Kids
*A Biblical Approach to
Earth Science and Earth History*

Contents

Materials list	5
How to Use this Study	7
Section I – Introduction to Biblical Geology	9
Part 1: The Creation of Earth	9
Part 2: Where Did All the Rocks Come From?	16
Part 3: What Do the Rocks Look Like?	20
Section II – The Two Contrasting Views of Earth History	27
Part 1: Telling a Different Story	27
Part 2: Two Approaches to Earth History	31
Section III – The Rocks and Minerals of the Earth	35
Part 1: Atoms and Elements – The Stuff of Rocks and Minerals	35
Part 2: Minerals – the Stuff of Rocks	40
Part 3: The Plutonic Rocks	44
Part 4: The Volcanic Lava Rocks	47
Part 5: The Pyroclastic Volcanic Rocks	52
Part 6: The Metamorphic Rocks	57
Part 7: The Sedimentary Rocks	62
Section IV – Fossils and Noah's Flood	73
Part 1: What is a Fossil?	73
Part 2: Types of Fossils	79
Part 3: Famous Fossils	83

Part 4: What do Fossils Really Tell Us?	89
Part 5: Where do We Find Fossils	95
Section V – Dinosaurs and Noah's Flood	100
Part 1: What are Dinosaurs?	100
Part 2: Kinds of Dinosaurs	104
Part 3: Were Dinosaurs on the Ark of Noah?	113
Section VI: The Oceans	122
Part 1: Ocean Features, Marine Fossils and Corals	120
Part 2: Ocean Rocks and Sand	134
Appendix A Fossil Card Template	139
Appendix B Rock Groupings Charts	140
Picture Credits	141

Materials List

If you have purchased the complete kit for *Geology for Kids*, you have the following materials in your kit. If you have only purchased the book, you will need to acquire the following specimens to fully illustrate the concepts taught in the book.

1. The following rock-forming minerals: quartz, potassium feldspar, muscovite mica, sodium feldspar, jasper, calcite, calcium feldspar, pyroxene, amphibole, magnetite, biotite mica, olivine
2. The following rock specimens illustrating the various rock types: granite, gabbro, basalt, rhyolite, ash, tuff, volcanic bomb, cinders gneiss, schist, quartzite, marble, shale, chert, siltstone, sandstone, conglomerate, breccia, limestone, halite, travertine, fossil limestone, chalk, bituminous coal
3. The following fossils: graveyard fossil, fern, clam, coral, marine reptile fossil, agatized wood, fossil cast/mold, worm tubes, gastrolith, gastropod, sea urchin, ammonite, trilobite, dinosaur bone, fossil mammal bone, leaf fossil, fossil algae, marine fossil from high in the mountains, otodus shark tooth, megalodon tooth, crinoid fossil, dugong bone, fossil fish, fossil fish vertebra
4. Oceanic fossils and specimens: kimberlite, diabase, peridotite, modern coral and sea shells, white sand, green sand, black sand, magnetic sand

You will also need the following materials for some of the activities:
- Bible
- Lab notebook
- Small plastic container (approximate size four in. by four in.)
- Plaster of Paris
- Vaseline
- Leaf/twig or shell
- Modeling clay
- Poster board, magazines (this project can also be done entirely on computer, if desired, so poster board and magazines would not be necessary)
- Two slices of bread
- Creamy spread (peanut butter, cream cheese, hummus, etc.)
- Sugar, brown or white
- Nuts, seeds (alternative could be dried cranberries, raisins, etc.)
- Banana slices
- Plate

How to use this study

1. Each section is divided into smaller parts that are easily manageable in one sitting. The first two sections are devoted to learning the scripture that stands behind the study. These sections are crucial as they set the stage for understanding Earth history and Earth science from a Biblical view. Don't gloss over these lessons! The sections after these will delve into the science of rocks, minerals, Earth forms and fossils.
2. We would suggest that your child keep a science notebook that is a combination of notes and labs. For the first two sections, there will be several Bible verses to memorize, and conversations that you will want to have with your child. So, for instance, for the verses, you might have your child copy the verses into the notebook. For the hands-on activities, you might have them record their observations in their notebook. Part of science is learning to keep a notebook with observations, drawings, notes, etc. This is a valuable discipline to learn. Observations can take the form of drawings or written notes.
3. Although your child may be able to do this study on his/her own, we highly recommend that you do the study together. It gives an excellent opportunity for your child to learn to express these concepts and have a dialogue over them with somebody. Read it together and talk about it.
4. There are 21 lessons in the book. We would suggest that you cover two of them a week so that not too much time passes before you come back to the subject.
5. I often pose questions to the student in the main part of the text. You might consider having your student answer those

questions in their notebook. Some are factual, but some would require them to think through the information that has been presented and draw conclusions.

6. Be sure to watch for some fun facts from *Rock-man Pat* as you go through the text. Some of these may provide a springboard for you to research a related area of geology.

Section I - An Introduction to Biblical Geology

Part 1: The Creation of Earth

Dear Parents, people who believe the Bible are ridiculed when it comes to things like the Creation and the Genesis Flood. I have seen many Christians compromise their trust in the Bible because of what people say. While we wrestle with our faith at times, think of what your child will go through when he/she is older. We must start now to teach them about how to view the earth from beginning to end. The Bible is more than a philosophy for life. Much of the Bible is recorded history supervised by the One who was there and saw it all. Teaching our kids to study the rocks in light of what God has spoken, will give them a solid foundation for trusting God's word in the future as they are challenged again and again. As you progress through the different learning levels of your kids, you can build off the material presented here. Kids are naturally drawn to rocks. Let's build off that interest by telling them the whole story – not just the Sunday morning portion of it.

There are a few things that a child of this age should be made aware of and that will serve as the foundation for the development of a Biblical view of Earth history. That is what is covered in the early grades. The following material should be covered in a one-toone dialogue with your child. You may feel free to cover this material at your and your child's pace of learning. Your child may have these things down or they may need to learn them. God bless you in your endeavor to teach your children the true and real history of the earth.

The word geology comes from two Greek words – *geo*, meaning earth, and *logi*, meaning the study of. So, geology is the study of the earth. When we study the earth, we should be curious about who made the earth, how and why it was made, all the rocks we see and the future of the earth. Aren't you a little curious about these things?

But, to answer these questions, we need to know someone who was actually there when the earth was made. Let's see. Who could that be?

Think back to your favorite birthday party. Were you there? Who else was there? What kinds of things did you do? *Make a list of the things you remember:*

_____

_____

The things you listed are called *evidences*. These are additional things that help me have trust in what you say.

Now, I wasn't there when you had that fun birthday party. How will I know about your birthday party? Well, you told me so. I believe you because you told me. And, I can also believe you because of the *evidences* you listed above. The same would be true of your parent's marriage. You know all about their marriage because they told you and because of the pictures and the reports from others who saw their marriage. Were you there? No, but others were. And so, you believe them.

The same is true when the earth was made. You weren't there. Your parents weren't there. Your friends weren't there. Your teachers weren't there. But God was there. And He has told us so in His Bible.

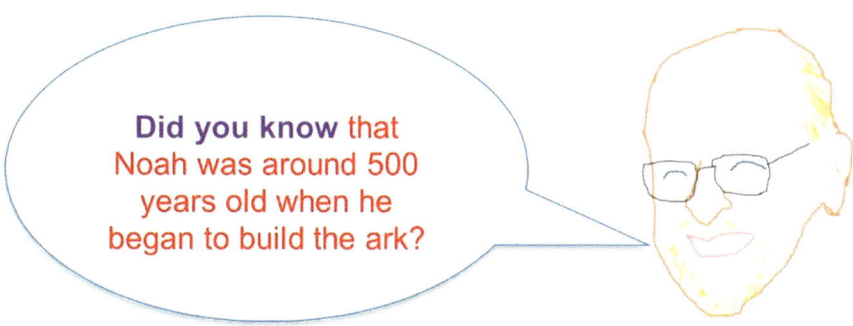

Activity
Memorization: *Genesis 1:1*

So, God was there when the earth was made and He has told me so in His Bible. Here is a great little saying that I learned early on in my Christian life: *"God said it. I believe it. That settles it."* This is what faith is all about. Faith is trusting the word of the One who was actually there when the earth was made.

But just like you had evidence of your birthday party, so God has left us with evidence that He made the earth. *Can you think of evidence that God has left us that He made the earth?*

_____
_____

Now that we have begun a good habit of listening to what God said about how the earth was made, do you think God has anything to say about the rocks that we see around us?

Of course, God's main interest is in people, isn't it? How do we know this? Well, God's Bible tells me so.

Activity

Memorization: *John 3:16*

OK. So, where did all the rocks come from? Remember, rocks are part of the evidence that God left us about what He did when He made the earth. Some rocks are pretty and they show the work of God Who is a Master Artist. Some are useful to us and therefore show God's care for us. Some rocks are just interesting, like God is. I think rocks from volcanoes are especially interesting. *What are some of your favorite rocks?*

_____
_____

So, some rocks show something about our God as our Maker. But some rocks, although interesting to look at, seem to tell a different story. Some rocks are twisted and crushed. Can you find a rock that looks like this?

Can you think of a story from God's Bible that might tell us how these rocks got this way?

_____

_____

If you guessed the story of Noah's Flood, you would be correct. Let's get God's Bible and read about Noah and his life from *Genesis chapters 6-8*. Now, remember, God was there when the earth was flooded. So, we can trust what He says about this event. *When you have finished listening to the story, list all the phrases that might tell us something about what happened to the earth during Noah's life.*

_____

_____

_____

So, some of the rocks we see around us will definitely tell us a story about God, our Maker. But, most of the rocks around us will tell the story of how God destroyed much of the earth He had made. *Why did He do this?*

_____

Activity

Memorization: *Genesis 6:5*

To me, this is one of the saddest verses of God's Bible. God made a beautiful Earth, but because of the evil of man, God had to destroy it. And much of the earth we see today is evidence of God's judgment.

The rocks today tell us that God is watching man and will hold him responsible for what he does.

Even most of the dinosaurs died as a result of man's evil. How do we know this? Because the fossils are evidence of this. Fossils are the remains of once living plants and animals preserved in the rocks.

Have you ever found a fossil? Fossils tell us how powerful the Flood of Genesis was. This was no ordinary flood! This Flood had never happened before and it has not happened since. Most fossils I have found tell me a story that matches what I read in God's Bible.

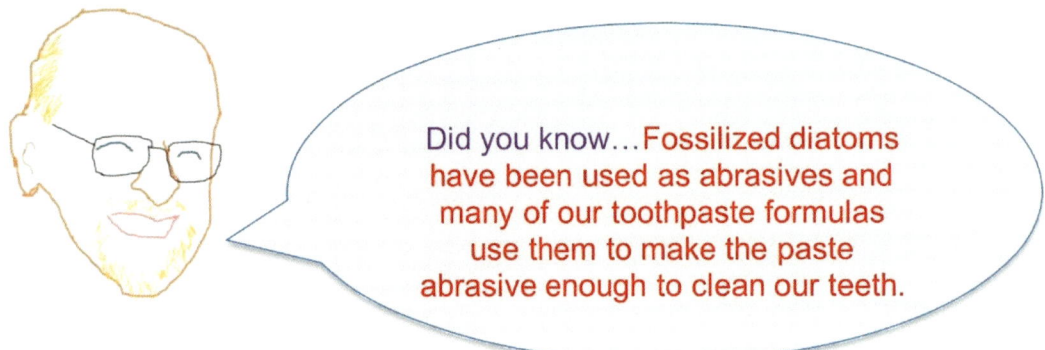

Did you know...Fossilized diatoms have been used as abrasives and many of our toothpaste formulas use them to make the paste abrasive enough to clean our teeth.

Activity
Memorization: *Genesis 7:23*

Another piece of evidence about the Flood that we can see today is found in all the volcanoes that are around the earth. Some volcanoes have stopped erupting and only their mountains are left. But some are still erupting. Where did all the volcanoes come from? I think I know. Let's see if you can guess? Read this verse from God's Bible.

Activity

Memorization: *Genesis 7:11*

So, it just didn't rain for forty days and nights, but the whole Earth was torn apart. Huge earthquakes must have taken place. Many mountains must have crumbled. Tremendous amounts of magma (lava) must have shot up from below the earth to form volcanoes. *Can you think of other things that must have happened with so much earth movement?*

_____

_____

All around the earth we can still see what has been left of the Flood. The earth is weak in many places. Earthquakes, tsunamis, tornados, volcanic eruptions are all a part of the weakened Earth as a result of the Flood. God helps us when these things affect us, but He has left them as evidences of the tremendous Flood that once destroyed the earth.

This is Biblical Geology! The rocks all around us tell a story of destruction and awesome power, like what God's Bible tells me. Some people may want to tell a different story, but they weren't there, were they? Who would you rather trust? God Who was there, or the people who were not there, but tell a different story?

Section I - An Introduction to Biblical Geology

Part 2: Where Did All the Rocks Come From?

Remember, rocks are part of the evidence that God left us about what He did when He made the earth. Some rocks are pleasant to look at and they show the work of God Who is a Master Artist. Some are useful to us and therefore show God's care for us. Some rocks are just interesting, like God is. I think rocks from volcanoes are especially interesting.

What are some of your favorite rocks?

_____

_____

Some rocks show something about our God as our Maker. But some rocks, although interesting to look at, seem to tell a different story. Some rocks are twisted and crushed. Can you find a rock that looks like this?

Most of the rocks around us will tell the story of how God destroyed much of the earth He had made. Why did He do this?
_____
_____

Activity
Memorization: Genesis 6:5

In the first section, we saw that much of the earth we see today is evidence of God's judgment. Only a remnant of what God originally made is reflected. The rocks today tell us that God is watching man and will hold him responsible for what he does.

Please note: In this study, I use these four colors consistently to help identify the different rock types.

<h3 style="text-align:center">Plutonic   Volcanic   Metamorphic   Sedimentary</h3>

*Many group the rocks into just three groups, combining the plutonic and the volcanic into one group, called igneous. (The word **igneous** means **fire**.) But this presumes that these two were both made by fire or from molten rock. Volcanic rocks are the only ones we have seen form that way. Consequently, I prefer to categorize the igneous rocks into two distinct groups.*

All rocks can be classified into four basic rock types:
- **Plutonic rocks** – those that make up the foundation of the earth. No one has ever seen these rocks form, so we don't know exactly how they were formed. We can study their chemistry and the minerals in them, but we don't know exactly how they came about.
- **Volcanic rocks** – those made from fire. We have seen plenty of these forming.

One of the many eruptions of basalt lava at Kilauea on the Big Island of Hawaii

- **Metamorphic rocks** – existing rocks that appear to have been twisted and changed. No one has ever seen these rocks form. Geologists think that they were formed through heat and pressure.

- **Sedimentary rocks** – those laid down by water, mud and other sediments. To a certain degree we have seen these forming, but not to the extent of the many sedimentary mountains and formations we see around the world.

When we study minerals, I use **light blue** and **dark blue** for the **light-colored minerals** and the **dark-colored minerals**. Be careful to not confuse this with the **plutonic rocks**. You should be able to tell which is which by the context of the study, whether I am talking about minerals or rocks.

Why do modern geologists group plutonic and volcanic rocks together into igneous rocks? Modern geologists group rocks according to how they *think* they formed. How did they develop this thinking? Over 200 years ago, geologists decided that they would no longer believe in Noah's Flood. The only other way to think of how rocks came to be was that they must have formed over millions and millions of years.

Activity
Memorization: *Genesis 7:11*

Section I - An Introduction to Biblical Geology

Part 3: What Do the Rocks Look Like?

We will go into the specifics of the rock types later. For now, let's just enjoy the obvious differences that we see in them.

Plutonic Rocks

Volcanic Rocks

Metamorphic Rocks

Sedimentary Rocks

There must have been a great amount of earth movement that took place during Noah's Flood. This event might explain why we see huge amounts of the foundation rocks like granite on the surface of the earth.

It is difficult to imagine such huge amounts of granite like that found in Yosemite National Park.

Another piece of evidence about the Flood that we can see today is found in all the volcanoes that are around the earth. Some volcanoes have stopped erupting and only their mountains are left. But some are still erupting. Where did all the volcanoes come from? I think I know. Let's see if you can guess? Remember your verse from God's Bible that you memorized, Genesis 7:11?

Mt. Adams in southern Washington towering over the city of Yakima at 12,281 feet. It is a massive active volcano that was probably initially formed toward the end of the Genesis Flood and covered in ice during the post-Flood ice age.

Agathla Peak in northern Arizona with its imposing 1,500-foot structure over the landscape. It is considered to be the eroded neck of an extinct magma intrusion. That is, it never quite erupted through the sediments that had covered it. Where are the sediments? They were probably washed away during the receding stage of the Genesis Flood.

When the fountains of the great deep burst open, along with lots of water that must have poured through the cracks in the earth, there must also have been a lot of volcanic eruptions. One such volcanic eruption might have been the Absaroka Mountains, remnants of a huge volcanic event. This range of volcanic mountains covers portions of southern Montana and the east side of Yellowstone Park in Wyoming. Geologists estimate that there was over 9,200 cubic miles of volcanic material ejected from these volcanoes! How big is that? It is such a huge number that we can only compare it with something else. Everyone considers the eruption of Mt. St. Helens in Washington State to have been a big volcanic eruption. But when Mt. St. Helens erupted in 1980, it ejected only about .25 cubic miles of volcanic material. The Absaroka Mountains are made of hundreds of times as much! That must have been a huge event!

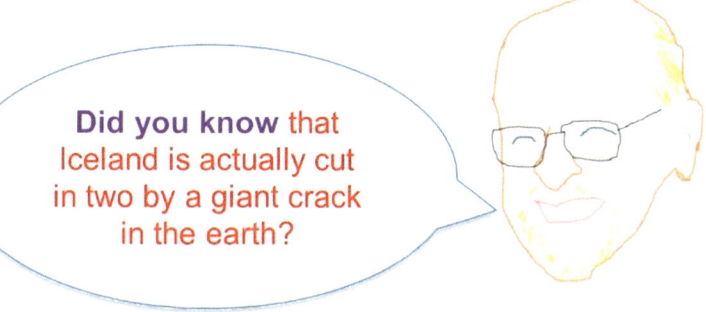

Glacially carved remnants of the volcanic Absaroka Mountains in Montana and Wyoming

As a result of the earth separating on what is now the ocean floor, tremendous amounts of steam and lava would have been released causing major changes in our atmosphere. The oceans left from the Flood would have been warmed releasing much moisture into the air. This is a recipe for an *Ice Age*. Thousands of volcanoes were probably erupting during and after the Flood. The combination of warmer oceans and a cooler atmosphere would have caused a great deal of snow and ice to form. Preserved on the earth today is evidence of this massive amount of ice and snow that came because of this *Ice Age*. Below is a picture of what is called a glacial valley. It is in the Beartooth Mountains of Montana. There is no more ice today, but the evidence of the ice that once existed is there!

The huge glacial valley pictured in this photo is several thousand feet long and deep! The Beartooth Mountains in Montana border Yellowstone National Park and bear evidence of having been completely covered by ice at one time in the past.

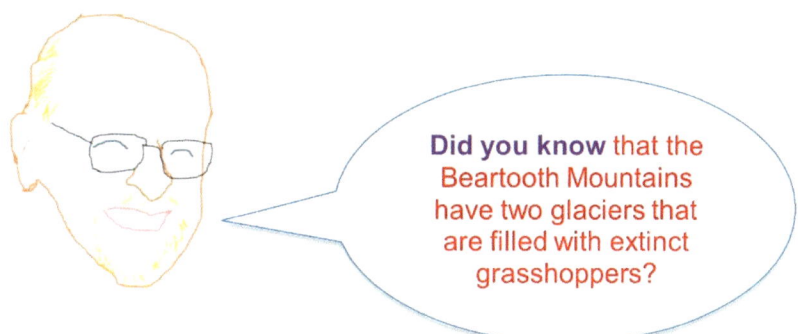

Did you know that the Beartooth Mountains have two glaciers that are filled with extinct grasshoppers?

During Noah's Flood, huge chunks of earth split apart and separated. The heat and pressure generated from this must have been tremendous. Rub your hands together. Do you feel the heat?

Imagine now, huge pieces of continents and mountains moving against each other. Metamorphic rocks we see today might have been a direct result of a tremendous amount of this kind of heat and energy.

Mount Shuksan, Cascade Range, Washington State

The Flood was God's judgment on the earth as a result of man's sin. Billions of living things died in this massive global flood! The sedimentary rocks contain remnants of these past creatures in the form of fossils. Even most of the dinosaurs died because of man's evil. How do we know this? Because the fossils are evidence of this. Fossils are the remains of once living plants and animals preserved in the rocks. Have you ever found a fossil? Fossils tell us how powerful the Flood of Genesis was. This was no ordinary flood! This Flood had never happened before and it has not happened since. Most fossils I have found tell me a story that matches what I read in God's Bible.

Fossil graveyards

Activity

Memorization: *Genesis 7:23*

So, it just didn't rain for forty days and nights, but the whole earth was torn apart. Huge earthquakes must have taken place. Many mountains must have crumbled. Tremendous amounts of magma (lava) must have shot up from below the earth to form volcanoes. With huge amounts of volcanic ash being hurled into the air, what kind of consequences would this have had on our earth?

_____

_____

This is Biblical Geology! The rocks all around us tell a story of destruction and awesome power, like what God's Bible tells us. Some people may want to tell a different story, but they weren't there, were they? Who would you rather trust? God Who was there, or the people who were not there, but tell a different story?

Activity

A valuable skill to learn as a scientist is to learn to make observations and record what you see. Take out the rocks in your kit and classify them according to size. Then classify them again according to color. Are they light/dark? Red/brown/green/black?
How else could you classify them?

Section II – The Two Contrasting Views of Earth History

Part 1: Telling a Different Story

Why did people start telling a different story from that told in the Bible?

Modern Geology had its beginning just a short time ago – in the early 1800s. As historical movements usually do, modern geology began with the influence of a few men; men who distrusted the Bible, its miracles and its history. A movement called The Enlightenment influenced these men to think differently. The Enlightenment was very similar to challenges that most humans experience when they are young. And some never change from that view. The men of The Enlightenment thought that they knew more than the Bible and proceeded to say so. Maybe you are struggling with this same feeling. You might think you know more than your

parents. But God gave us people in our lives that are there to help us so that we don't fall into false ideas. The result of this rebellion was a world in which God was not welcome and was thought to be irrelevant. One such influential man was James Hutton, a Scottish scientist.

James Hutton wrote extensively in the late 1700s that the history of the earth must be interpreted solely by naturalistic causes and reasoning. In other words, he wanted nothing to do with God's historical document, the Bible. Why would he say this?

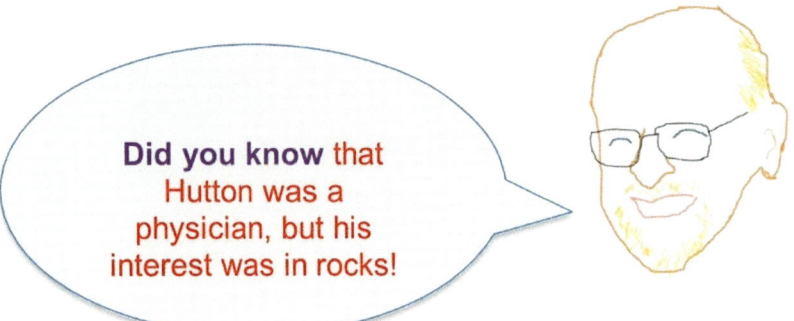

Many people have overlooked the culture in the time of Hutton and others. Western Civilization was in the midst of a philosophical movement called, The Enlightenment or The Age of Reason. While many historians see this time as a sort of liberation for mankind, for the church it was the beginning of a dark period. We are still suffering the effects of it today. The liberation that some historians talk about has to do with a sort of overthrow of God in geology and ultimately in the rest of science. A similar sort of thing will happen in the last days, only with much more severe consequences.

Activity

Memorization: *Psalm 2:1-3*. What do these verses have to do with our study of geology?

There were many geologists who spoke out against the *New Geology* as it was called. But more and more geologists, many of who were pastors and theologians began to compromise the Scriptures. Slowly but surely compromise gave way to rejection of the Biblical history contained in Genesis.

The next major shift came in 1830 when Sir Charles Lyell published his Principles of Geology. This work would be the predominant influence in another major player in the new science – Charles Darwin.

Charles Lyell   Charles Darwin

Publicly, Lyell appeared to be scientific and reasonable. Privately he wanted to eradicate any reference to Moses in geology. That included of course – the Genesis Flood. So, by the late 1800s there were no scientists who believed or even referred to the Genesis Flood. Lyell had been victorious – at least for the next hundred years!

Then in 1961 a book was published that would begin to level the playing field. That book, *The Genesis Flood*, by the late Henry Morris and John Whitcomb, started sort of a revolution in geological thinking.

Today, Dr. Morris' book has radically influenced many hundreds of people and scientists, including myself. We owe a debt of gratitude to Dr. Morris for his courageous stand. You see, he was a geologist who risked his career to publish his research. He serves as an example for any who would dare speak out against the *New Geology*.

Why would people want to tell a different story from that of the Scriptures? Realize that the debate over Scripture and Science is really not about science but about rebelling against our Maker. The rocks themselves do not have writing on them. They do not have dates published on them. The rocks must be *interpreted;* and interpreted by people who have chosen to believe certain things about what they see. Motives are not always apparent to us. But they are to God. What people really think is often masked by supposed scientific reasoning. It is interesting to note that Charles Lyell was one way in public and in his intimate letters to his supporters revealed his true self. How we view the world around us is called our worldview. Let's make sure our worldview is based on the Scriptures. After all the Scriptures are the words of the One who was there in the beginning, at the Flood and in Christ 2000 years ago. And this One can be trusted. He has not done anything that would lose our respect and trust.

This is Biblical Geology! It is a totally different approach to the study of Earth History and Earth Science. In the following sections, you will learn not only the science of the earth, but also its origin and the reasons for the rocks that we see all around us.

Activity
Write a short report on either Lyell or Darwin. Answer questions such as: When was he born? When did he die? Where did he live? What was his schooling? What are some of the important things that he did during his life?

Section II – The Two Contrasting Views of Earth History

Part 2: Two Approaches to Earth History

Read 2 Peter 3:3-8.

What do these verses say about the creation of the world? What do you notice about what the Lord says will happen on the Day of Judgment?

This passage of Scripture also seems almost prophetic as it describes a philosophy about the earth that will become prevalent among mankind. (The word *philosophy* can be described as your view of life.) That philosophy is *naturalism*. It is the belief that the history of the earth has been going on for a long, long time with nothing different than what has always been observed. Does this sound familiar? It should, because it is describing a belief called *uniformitarianism*. This means that all of Earth's geological and biological processes have been going on as naturalistic processes without the interference of a god. It is the logical conclusion when God is rejected.

Uniformitarianism is often defined this way: "The present is the key to the past." What do you think this means?

Activity

Below is a chart comparing the two approaches to Earth History. Take some time and discuss it with another person. Draw two different series of pictures to illustrate the major differences. (For instance, the Biblical series might be a picture for each day of the creation week.)

A Comparison of the Two Approaches to Understanding Earth History:

Genesis – Biblical Geology	Uniformitarianism – Secular Geology
Theism (God, but a specific God, the God of the Scriptures) – God has been actively involved in the origin and history of the earth from the beginning; we call God's involvement in His creation, *miracles*.	Deism (the deity; not identified, but sometimes referred to as *the Supreme Architect*) – the Deity may have created the earth and universe but if He did, He is no longer involved in it; He does not intervene. Only natural laws govern the history of the earth; there are no such things as miracles
Gen. 1:1 – **in the beginning,** *God; beginning* **would mean of time and of matter; God is eternal; He did not have a beginning. Another way to say this is:** *in the beginning of our cosmos,* **God was already present. And it is He who has given order and meaning to creation.**	Big Bang – in the beginning *matter and space* were already there; the Big Bang does not attempt to answer the question, "Where did matter and space come from?" It is ignored.

Gen. 1:1 – God created the *space* and the *earth* on Day One of the creation (created means, *out of nothing*); Earth was the only *heavenly body* in the beginning.	Matter exploded, forming galaxies, stars, and planets 15 billion years ago; Earth came along about 4.6 billion years ago, or 11 billion years after the universe came into existence.
Gen. 1:2 – the earth was originally a surging mass of *water* (**the deep**).	The earth was originally a cloud of gas or a molten blob of magma/lava; water came last
Gen. 1:2 – **the** Spirit of God was an active part of creation from the beginning.	God, spirit and religion are not matter and consequently are not relevant and are excluded from the study of Earth history
Gen. 1:3 – *light* (energy; the physical laws that govern the universe)	Physical laws and energy *evolved* as the universe evolved

Gen. 1:5 – *good* = perfect, done, complete in God's eyes; evening and morning define a complete revolution of the earth = day; *one day* means there was one complete 24-hour period of Earth history with evening and morning (time divisions) and it was the first day of more to come	Day in the Scriptures is simply *symbolical* for a period of time; because of the undeveloped minds of these early people, cosmology had to be communicated simply; they could not comprehend *deep time*; according to evolution, nothing is ever completed, but always changing into the next thing over millions of years (*deep time*)

When man rejects God's explanation of the origin of the earth, then man is left with an explanation that does not make sense in light of what we see and know about the earth. The earth is a wonderful place, uniquely made and uniquely placed in a huge lifeless universe. This is exactly what David concluded in the Psalm 139 - *I am fearfully and wonderfully made.*

Section III – The Rocks and Minerals of the Earth

Part 1: Atoms and Elements – The Stuff of Rocks and Minerals

What is a rock? What is a mineral? What are they made of? They capture our attention. Kids as well as adults are fascinated with rocks and minerals.

A rock is made of minerals. For example, the rock granite is made of three minerals – quartz, feldspar and mica.

What is a mineral? A mineral is made of elements, which are made of atoms, which cannot be seen without the help of powerful microscopes. In order to see an atom, scientists have drawn pictures to help teach what an atom might look like. The word atom is a Greek word. The ancient Greeks thought that material things were made of things that could not be seen. They called these atoms, even though they did not know exactly what they were. The word *atom* means, *without cutting*. So, the Greeks thought that material things must be made of things that could not be broken down into anything smaller. The Greeks were mostly right. Today scientists draw pictures that look like this:

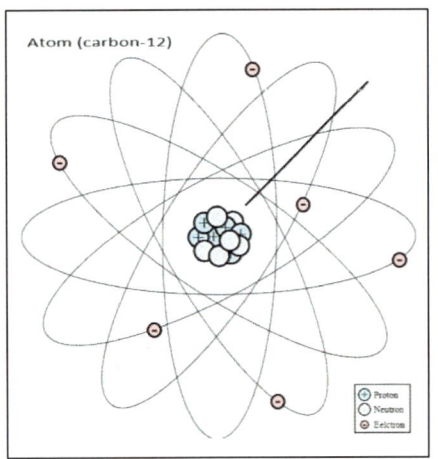 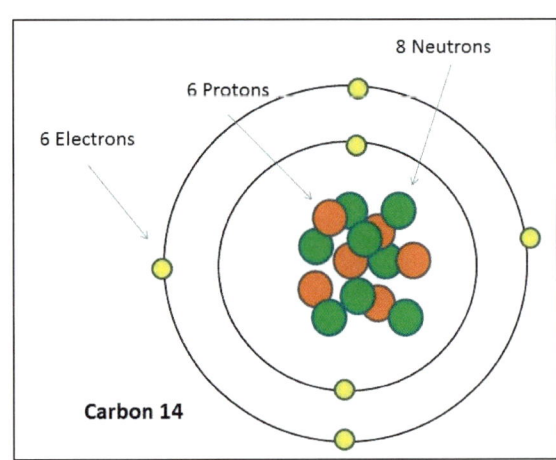

An atom has three parts that you need to know:
1. **The electron** – a very small part of the atom that has an electrical charge that we will call, negative (-)
2. **The proton** – a larger part of the atom that has an electrical charge that we will call, positive (+)
3. **The neutron** – a part of the atom that is grouped with the proton that has a neutral electrical charge.

Now, you might say that the atom exists because of these charges negative, positive and neutral. But that does not answer the question. Why do these atoms exist in the first place? And what KEEPS them together? These are questions that science cannot answer. But there is one source that can answer that question – the

Bible. Read the passage of the Bible found in the New Testament letter to the Hebrews, chapter one and verse 3. This passage tells us that it is the Son of God (Jesus) who holds all
things together by the word of his power. Sure, it is God who keeps things working like He created them to do.

Atoms are specially designed by God to form things called elements. These elements are the stuff of minerals. Elements are very important as they make up the air we breathe, the food we eat and the sunshine we enjoy.

The earth is a very special place designed and created by God for man, His special creation, to live. The elements that make up the air we breathe and the dirt and rocks we walk on are all specially designed by God to make a suitable home for man. There are eight elements that are the most abundant elements in the earth's crust:

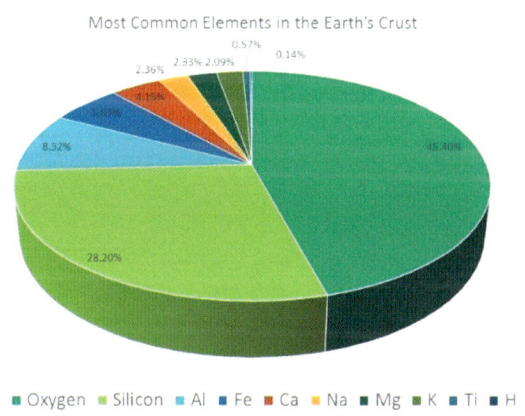

1. Oxygen 46.4%
2. Silicon 28.2%
3. Aluminum 8.32%
4. Iron 5.63%
5. Calcium 4.15%
6. Sodium 2.36%
7. Magnesium 2.33%
8. Potassium 2.09%
9. Titanium 0.57%
10. Hydrogen 0.14%

I bet you have seen these elements on the back of the cereal box you read while having breakfast one morning. There are 118 elements that have been discovered so far. The elements in the picture above are all combined in very special ways to make minerals. These eight elements form the basic minerals that form most of the rocks. For example, the mineral quartz is made of two elements, silicon and oxygen. We call these minerals, The Rock- forming Minerals. You will learn about these later.

What do we really know about the earth?
Geologists tell us that the earth is made up of at least **six** parts. Geologists call these:
1. **The continental crust (made up primarily of** granite**)** ✓
2. **The oceanic crust (made up primarily of** basalt**) Together these** ✓ **are known as the** *lithosphere* **or** *rock sphere*
3. **The** *asthenosphere* **or** *weak sphere* **(thought to be somewhat fluid viscous material; sometimes called the** *upper mantle***). Viscosity is a term that measures the fluidity of something like water and honey.**
4. **The mantle (thought to be solid material sometimes called the** *lower mantle***)**
5. **The outer core (thought to be liquid)** ✓
6. **The inner core (thought to be solid)** ✓

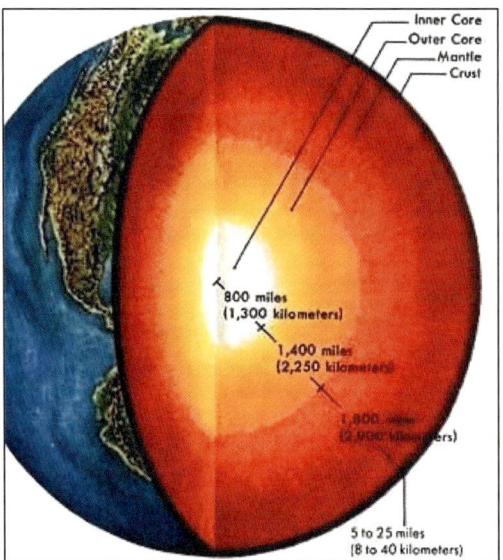

Is this known for sure? This may surprise you, but it is not known exactly what the interior of the earth looks like or what it is made of. We know many of the minerals involved and we are able to study at least some of the continental crust and oceanic crust. The rest of it is a guess. Probably the most important thing for you to remember is found in the New Testament in the letter to the Hebrews. **Read Hebrews 1:1-3.** It is Christ Himself who holds the world together and keeps it operating. We may discover the exact makeup of the earth's interior someday, but for now it is comforting and reassuring to know who it is that really holds things together.

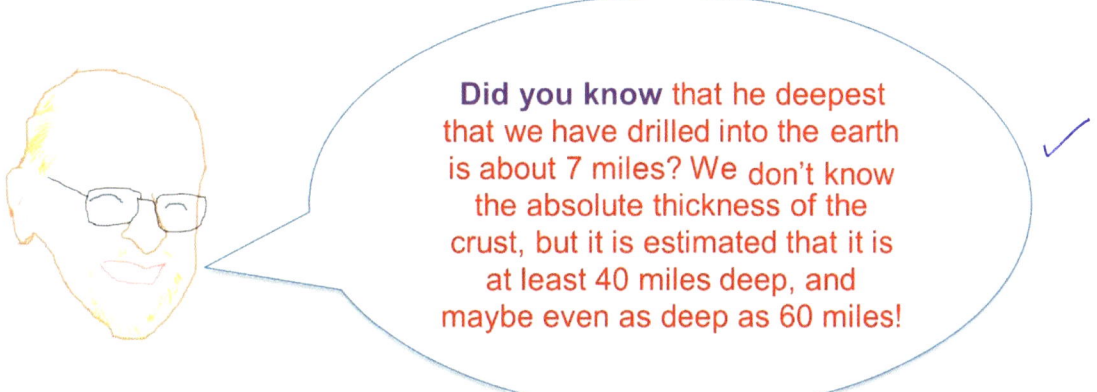

Did you know that he deepest that we have drilled into the earth is about 7 miles? We don't know the absolute thickness of the crust, but it is estimated that it is at least 40 miles deep, and maybe even as deep as 60 miles!

Review

1. What is an atom?
2. What is an element?
3. What are the parts of an atom?
4. What is a mineral?
5. What is a rock?

Write your answers to these questions in your notebook.

Activity

Using a dictionary or Internet resources, look up the eight most abundant elements in the earth and make a chart showing how many electrons, protons and neutrons each one has.

Quiz

1. What is a rock?
2. What is a mineral?
3. What are the three parts of an atom?
4. What are the two parts to the crust of the earth, and what are they made of

Section III – The Rocks and Minerals of the Earth

Part 2: Minerals – the Stuff of Rocks

In our last lesson, we discovered that everything is made of atoms. These atoms are specially designed by God to form elements. Elements are specially designed by God to form minerals. Minerals form rocks. An example of a mineral is quartz. An example of a rock is granite.

Quartz is a mineral.   **Granite is a rock.**

Did you know that there are over 4,000 minerals that have been identified by scientists? It would take many years to learn all the minerals on Earth. But to have fun with the rocks in your back yard, you only have to learn 12!

These 12 minerals are responsible for forming most of the rocks we collect. Here they are.

1. Quartz
2. Potassium feldspar
3. Muscovite mica
4. Sodium feldspar
5. Jasper (a colored type of quartz)
6. Calcite
7. Calcium feldspar
8. Pyroxene (pronounced as PEER-ux-seen)
9. Amphibole (pronounce as AM-fi-bole)
10. Magnetite (a form of iron)
11. Biotite mica
12. Olivine (pronounced as AH-li-veen); a form of magnesium

All these minerals are formed from the eight most abundant elements that God created to make up the very special place we call Earth.

If you can learn the names of these 12 minerals and can identify them, you are well on your way to identifying almost any rock in the field!

These rock-forming minerals can be divided according to light and dark minerals. The first six in your list above are called light- colored rock-forming minerals. The next six in your list above are called dark-colored rock-forming minerals. These light and dark minerals will help you organize your rocks into light and dark-colored rocks.

The Light-Colored Minerals	The Dark-Colored Minerals
Quartz Jasper (type of quartz) Potassium feldspar Muscovite mica Sodium feldspar Calcite (typically associated with sedimentary rocks)	Olivine Amphibole (Hornblende) Pyroxene (Augite) Biotite mica Calcium feldspar Magnetite (iron)

The rock-forming minerals look like this:

They are: top row (left to right): *quartz, jasper, potassium feldspar, sodium feldspar;* **second row:** *calcium feldspar, biotite mica, muscovite mica, olivine;* **third row:** *pyroxene, amphibole, magnetite (iron);* **fourth row:** *calcite*

Review

1. What are the eight most abundant elements that form the earth's crust? You need to memorize these.
2. What are the 12 rock-forming minerals? You need to memorize these.

Write you answers in your notebook.

Activities

1. Using an encyclopedia or Internet resources, look up each of the 12 rock-forming minerals and make a chart listing the elements that make them up. What do you notice about the elements that make up the rock-forming minerals?

2. Take out the bag of rock-forming minerals from your kit and look at each one closely.

3. Organize the minerals into two groups: light-colored minerals and dark-colored minerals. Review these minerals and their two groups until you can recite them without looking at their labels. This activity is very important and it will help you understand rocks.

Quiz

1. Write out the 12 rock-forming minerals.
2. Name the eight most common elements of the earth's crust.

Section III – The Rocks and Minerals of the Earth

Part 3: The Plutonic Rocks

Plutonic Rocks	
Granite	Gabbro

Rocks are fascinating! They are so fun to collect. Mom has a hard time keeping her house clean because of them! Why are rocks so attractive to kids? I remember that my interest in rocks began when I was in second grade. I have had an interest in rocks ever since then.

What are rocks? Rocks are made of minerals. And there are just 12 minerals that you have to learn about in order to have a lot of fun collecting rocks. These 12 minerals are called rock-forming minerals. They give rocks their distinctive colors and make them light and dark.

Rocks can be divided into four special groups.
The Plutonic Family: Let's think about these groups as families. So, the first family you will learn about is made of the Plutonic rocks – named after the mythological god of the underworld, Pluto. The underworld was used to describe everything we cannot see below the earth's surface. These rocks form what we call the *basement rocks* or the *foundation rocks*. These rocks form most of the earth's crust. No one has ever seen a plutonic rock form. We have to guess where they came from. Because I believe the Bible story about the creation, I think that most of these plutonic rocks were created by God to form the foundation of the earth.

They are tough and full of minerals that can be easily seen with the naked eye.

Plutonic rocks can easily be identified by their large mineral crystals.

The plutonic rocks are divided into light-colored and dark-colored plutonic rocks. There are lots of plutonic rocks. But for now, we are going to concentrate on two types. As we go through the study, I am going to give the rocks a name to help you remember them. Some of the names will be evident, some will not, and I will explain those to you. The names are just an aid to help you memorize them. So, two plutonic family members that we are going to meet right now are 'Grainy' Granite and 'Gritty' Gabbro.

'Grainy' Granite wears light-colored suits. Do you see the *grains* or minerals? The material that makes up his suits is quartz and potassium feldspar with a little bit of biotite mica sprinkled in. Sometimes Grainy Granite will wear suits that have some muscovite mica too.

Granite

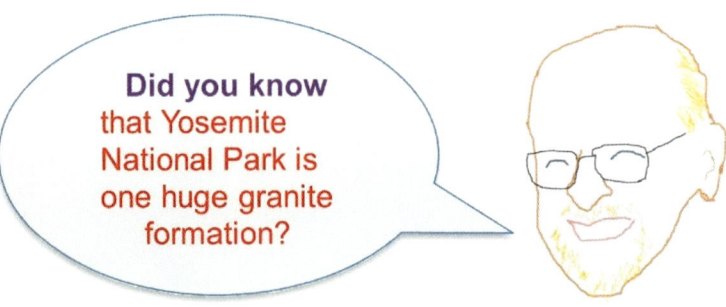

Did you know that Yosemite National Park is one huge granite formation?

Activity

Take out the samples of granite from your kit. How would you describe them? Draw a picture of them to show what you see. Record your observations/pictures in your notebook.

Now, pick out the minerals from your kit that match the minerals that make up the suits that Grainy Granite wears.

Now, take out the samples of gabbro in your kit. 'Gritty' Gabbro likes to wear dark-colored suits. *Gritty* can mean *rough*. Do you see the rough texture of the gabbro? The material that makes up his suit pyroxene, calcium feldspar, iron and sometimes calcium feldspar and amphibole sprinkled in.

Gabbro

Activity

Take out the samples of gabbro from your kit. How would you describe them? Record your observations in your notebook, using pictures if you like. Next, pick out the minerals from your kit that match the minerals that make up the suits that Gritty Gabbro wears.

In our next lesson, you will learn about volcanic rocks.

Section III – The Rocks and Minerals of the Earth

Part 4: The Volcanic Lava Rocks

Volcanic Lava Rocks	Pyroclastic Volcanic Rocks
Rhyolite	Ash
Basalt	Tuff
	Bombs
	Cinders

The volcanic lava rocks are the most interesting to me. They have more varieties of suits in their closet than I have ever seen! Scientists have seen many volcanic rocks form, but they have not seen where the material comes from that makes up their suits.

Types of Volcanoes

Shield volcanoes – are called *shield volcanoes* because of their shape. They look like upside down shields. They make shields because their lava is basalt which spreads out rapidly. One of the best examples of a shield volcano is Mauna Loa on the island of Hawaii.

Mauna Loa

Stratovolcanoes – are called stratovolcanoes because of the massive amounts of lava and volcanic rocks that are built up in layer after layer of eruptions. They form massive high cones and dominate the landscape. They are very easy to recognize. A great example of a stratovolcano is Mt. Shasta in northern California.

Mt. Shasta, California towers over the countryside at 14,179 feet!

Caldera – is Spanish for *big pot* from which we get caldron. Calderas are huge cavities in the earth that have erupted at some time in Earth history producing violent eruptions and great amounts of ash. One of the best examples of a caldera is Crater Lake in Oregon.

Crater Lake

Activity

Using modeling clay, make models of the three types of volcanoes.

The Volcanic Rocks Family: Within the volcanic rocks family there are two groups: the lavas and the pyroclastic rocks. The first group we will look at are the lavas. Just like the plutonic family, the basic volcanic lava rocks can be organized into light-colored and dark-colored volcanic rocks. (There are other volcanic lava rocks, but we will not cover them here.)

Scientists use two terms to describe the differences in molten rock. The molten material that is beneath ground is called *magma*. The molten material that comes up and onto the ground is called *lava*. Lava is thought to come up from deep within the earth and through pipes, passageways and fissures, including volcanoes.

Volcanic lava rocks can be easily recognized because you cannot see the individual mineral crystals. But you can easily see the dark color and the light color. Their names are 'Ruddy' Rhyolite and 'Blacky' Basalt.

Rhyolite

Rhyolite

Basalt

The word *ruddy* means reddish. Ruddy Rhyolite likes to wear light-colored suits. Sometimes Ruddy Rhyolite wears showy colorful suits! The material that makes up his suits is made of quartz and potassium feldspar with a little bit of amphibole sometimes sprinkled in. Ruddy Rhyolite's very colorful suits have a lot of iron mixed in. And many times, Ruddy Rhyolite likes to fool people with a dark suit called obsidian! It is made of quartz with a lot of iron mixed in to make people think it is a dark-colored rock!

Ruddy Rhyolite is a very thick lava because of the quartz that makes up his suits. He takes his time going from place to place, but never travels too far from his home.

Activity

Take out the samples of rhyolite from your kit. How would you describe them? Record your observations in your notebook, including any pictures you drew. Next, pick out the minerals from your kit that match the minerals that make up the suits that Ruddy Rhyolite wears.

'Blacky' Basalt likes to wear dark-colored suits. The material that makes up his suit is made of pyroxene, calcium feldspar, iron and sometimes sodium feldspar and amphibole sprinkled in.

Basalt

Blacky Basalt is thin lava because he has very little quartz that makes up his suits. He is usually in a hurry to go from place to place and likes to travel great distances from his home. An example of just how far Blacky Basalt can travel is the Columbia Plateau Basalt formation in Washington, Idaho and Oregon.

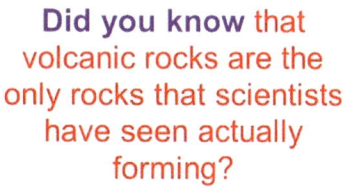 

Did you know that volcanic rocks are the only rocks that scientists have seen actually forming?

The Columbia Plateau is made up of over 60,000 square miles of basalt lava flows!

Activity

Take out the samples of basalt from your kit. How would you describe them? Record your observations in your notebook, including any pictures you drew. Next, pick out the minerals from your kit that match the minerals that make up the suits that Mr. Basalt wears.

In our next lesson, you will learn about the pyroclastic volcanic rocks.

Section III – The Rocks and Minerals of the Earth

Part 5: The Pyroclastic Volcanic Rocks

Volcanic Lava Rocks	Pyroclastic Volcanic Rocks
Rhyolite	Ash
Basalt	Tuff
	Bombs
	Cinders

The Pyroclastic Volcanic Rocks: The word *pyroclastic* means *fire broken*. They are formed from very hot volcanic explosions. Some of the pyroclastic volcanic rocks are thrown very great distances. Some like to pop and hiss close to their home. Like Ruddy Rhyolite and Blacky Basalt, the Pyroclastic family wears light- and dark-colored suits too.

The Pyroclastic group has lots of kids. The first kid is named 'Aerial' Ash. The word *aerial* means *airborne*. Ash is light-colored because of the amount of quartz and light-colored rocks it wears. Aerial Ash is made up of tiny bits and pieces of broken light-colored rocks and minerals. When Ash first enters the house, he makes quite an entrance! Study the photos below to notice how Ash erupts from volcanoes.

> Did you know that ash from the Yellowstone eruption has been found as far south as Louisiana? Locate these two places on a map.

Ash

Activity

Take out the ash and the light-colored minerals from your kit. Describe what you see. Record what you see in your notebook.

 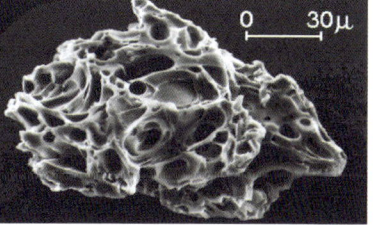

Ash and magnified ash

Ash that has built up into layers of rock

The next volcanic pyroclastic rock we want to introduce is the brother 'Tough' Tuff. The word tuff means, *porous rock*. That does not help very much because a lot of tuff is anything but a porous rock! Tough Tuff is made of bits and pieces of light-colored volcanic rocks and minerals fused together often into a hard rock. Tuff is actually ash and hot steam that is initially very hot when erupted. As it falls to the ground, it welds together into tuff.

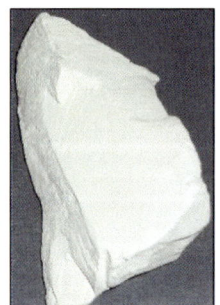
Tuff

Activity

Take out the samples of tuff and the light-colored minerals from your kit. Describe what you see. Record your observations, and any drawings you made, in your notebook.

Another pyroclastic rock is a dark-colored pyroclastic rock called 'Belching' Bomb. It is closely related to basalt, but did not flow out of a volcano, but was ejected (belched!) or expelled like a bomb shot out of a cannon!

Bombs

Activity

Take out the bomb sample from your kit and describe what you see. Do you have any idea why it is a rust color? Remember that basalt has a lot of iron in it. What does iron do when it is exposed to oxygen? It rusts and turns orange!

One other dark-colored pyroclastic rock to look at is 'Clinkery' Cinder. The word *clinker* is another name for cinder. Clinkery cinder comes from the same types of basalt eruptions and so are dark. The kind of volcano that produces cinders is called a cinder cone.

Cinders

A cinder cone – a type of volcano that erupts just once and in showers of hot basalt sparks.

Activity

Take out the samples of cinders from your kit and describe what you see. Record your observations, including pictures, in your notebook.

In our next lesson, you will learn about the Metamorphic Rocks.

Section III – The Rocks and Minerals of the Earth

Part 6: The Metamorphic Rocks

Foliated Metamorphic Rocks	Non-foliated Metamorphic Rocks
Gneiss	Quartzite
Schist	Marble

The Metamorphic Family: The third special group of rocks is made up of the metamorphic rocks. The metamorphic rocks can also be organized into light-colored and dark-colored metamorphic rocks.

Quartz, a light-colored mineral, will cause metamorphic rocks to be light-colored.

The word metamorphic is actually a Biblical word. It is found in the New Testament Letter to the Romans in chapter 12. Those who believe in Christ are commanded to be transformed (metamorphosed) into the image of Christ by changing their thinking about a lot of things. So, the word metamorphic means *change*.

Just how metamorphic rocks changed and what they were before they were changed is not totally understood by geologists today.

Why is that? Because no one has ever seen a metamorphic rock in the process of changing from one rock to another! Geologists guess as to where they came from.

Because I believe in the Bible as God's revelation about Earth history, I can see a clue to where metamorphic rocks came from. In Genesis chapter 7 we read about a global flood that came on the earth about 4,500 years ago.

Geologists think that two things were important in the formation of metamorphic rocks.
1. Lots of heat to melt rock
2. Lots of pressure to stretch and reshape rocks

So, if we think about the heat and pressure that were part of changing rocks as coming from the Genesis Flood, it becomes clear just how metamorphic rocks could have formed.

When the *fountains of the great deep burst open* in Genesis 7:11, huge amounts of existing foundation rock would have been moved. Much of this rock would have rubbed against other rock creating lots of heat and pressure. Layers of sediment that had been buried during the first days of the Flood would have been put under great amounts of pressure and heat that could have changed them into other rocks. We call the rocks that produced the metamorphic rocks, parent rocks.

There are two kinds of metamorphic rocks. The first kind is called foliated metamorphic rocks. Foliated comes from a Latin word meaning, *leafy*. These rocks are either layered or banded with alternating light and dark bands of minerals – the same light and dark minerals that formed plutonic rocks. We will look at *just two* of these foliated metamorphic rocks.

Foliated MetamorphicRocks

The foliated metamorphic rocks include:

'Stripy' Gneiss (Gneiss is pronounced like the word *nice*), and 'Shiny' Shist.

Activity

Take out the samples of gneiss from your kit and describe what you see. Record your observations and any pictures you drew in your notebook. What do you think the parent rock was? If you look at it closely, you will see that most likely the parent rock was granite. How many light and dark mineral bands can you see?

Gneiss

Now take out the sample of schist from your kit and look at it closely. Does it remind you of anything? Record your observations. To me it looks like the rock is covered with glitter. This is the light-colored mineral mica.

Schist

The second kind of metamorphic rock is called Non-Foliated. Non-Foliated metamorphic rocks are crystalline. They are filled with tiny mineral crystals that have been locked together. They remind me of a sugar cube. They do not show any layers. Let's look at two of these rocks.

The Non-Foliated metamorphic rocks include:
 'Quirky' Quartzite and
 'Monument'" Marble

Activity
'Quirky' Quartzite. *Quirky means peculiar.* Take out the samples of quartzite from your kit and look at it closely. Quirky Quartzite is indeed, peculiar. In the rough state, it can appear very ugly. When polished, it can be quite handsome. Quartzite has many colors of clothes, also, because of the different elements that can be in it. There is nothing secretive about Quirky Quartzite. Its name tells us that it is made of the light-colored mineral, quartz. Record your observations.

Quartzite

'Monument' Marble. Many monuments are made of marble. This rock also comes with many different changes of clothes! And many famous people have used marble to decorate their buildings with. Monument Marble has many disguises and you cannot tell for sure if what you have is marble until you drop a little bit of strong acid on it. It will fizz because the rock contains calcium carbonate, and it reacts to acid by fizzing. This is how you know that Monument Marble is made of the light-colored mineral, calcite (calcium carbonate). Take out the samples of marble from your kit and describe what you see. Record your observations.

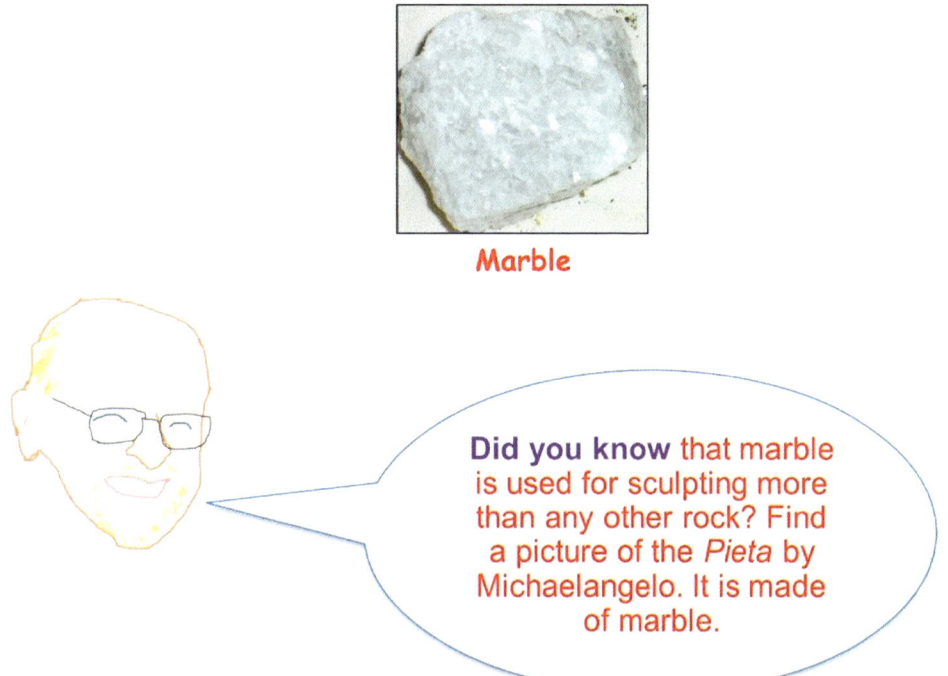

Marble

Did you know that marble is used for sculpting more than any other rock? Find a picture of the *Pieta* by Michaelangelo. It is made of marble.

In our next lesson, you will learn about the Sedimentary Rocks.

61

Section III – The Rocks and Minerals of the Earth

Part 7: The Sedimentary Rocks

Clastic Sedimentary Rocks	Chemical Sedimentary Rocks	Biochemical Sedimentary Rocks
Shale	Limestone (no fossils)	Fossil Limestone
Siltstone	Halite	Chalk
Sandstone	Travertine	Coal
Conglomerate		Chert
Breccia		

The fourth special group of rocks is made up of the sedimentary rocks. The sedimentary rocks are organized a little differently than the other three groups of rocks. The sedimentary rocks are grouped according to the bits and pieces of rocks and minerals that the sedimentary rocks wear.

The Sedimentary Rock Family: There are three brothers in the **sedimentary rock family.** They are **'Clunky' Clastic, 'Cunning' Chemical** and **'Biota' Biochemical.** Each one of the brothers has children in their family.

Most of the family is made up of the light-colored minerals, jasper, quartz and calcite.

The sedimentary rocks are very special because they, more than any other rock, record the great global flood of Noah's day. The word sedimentary is easy to remember because they are all made of sediments. Sediments made of mud and water. When the flood of Genesis came on the earth 4.500 years ago, it created a lot of mud! This mud came in different colors and sizes. As with the other families, there are many sedimentary rocks, so we will only examine a few of them.

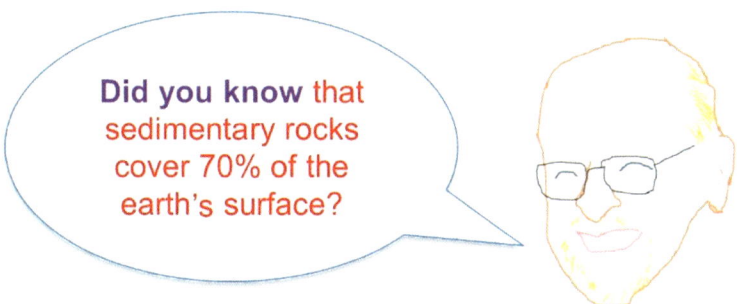

  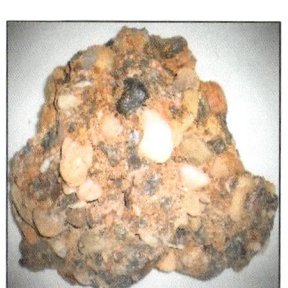

Sedimentary Rocks

'Clunky' Clastic's sedimentary rocks: I use the word 'clunky' to remind myself of the clumps of sediment that make up these rocks. Sedimentary rocks are self-organizing. That's a big word. It means that many mud sediments arrange themselves according to like sizes. Sedimentary rocks seem to naturally fall into lines according to the sizes of grains they contain. So, sedimentary rocks are organized according to the grain sizes. And the word clastic in Greek means *broken*. So, Clunky Clastic's kids are arranged according to the sizes of the *broken* bits and pieces of rocks and minerals they contain. Let's look at a few of these: **'Shaky' Shale, 'Soily' Siltstone, 'Sandy' Sandstone, 'Chunky' Conglomerate and 'Broken' Breccia.**

'Shaky' Shale. Shale is made up of very tiny bits and pieces of mud or clay. It tends to crumble easily, although some shale is hard.

Shale

Activity

Take out the shale from your kit and look at it closely. Describe what you see. Record your observations and/or drawings.

'Soily' Siltstone. Siltstone is made up of bits and pieces of clay and mud that are a little larger than shale. It doesn't have many uses other than as a contributor to soil, although sometimes it is used in sculptures and buildings.

Siltstone

Activity

Take out the siltstone from your kit and look at it closely. Describe what you see. Record your observations and/or drawings.

'Sandy' Sandstone. I have named this one *Sandy* because it is made of sand. That was an easy choice for a name! Sandstone is made up of bits and pieces of the light-colored mineral, quartz.

Sandstone; Quartz crystals magnified 200 times

Activity

Take out the sandstone from your kit and look at it closely. Describe what you see. Now look at it with a magnifying glass. What do you see? You should see very tiny quartz pebbles or tiny crystals. Record what you see.

'Chunky' Conglomerate. Conglomerate is a mix of rounded stones – mainly the light-colored mineral, jasper.

Conglomerate

Activity

Take out the conglomerate from your kit and look at it closely. Describe what you see. Record your observations and/or drawings.

'Broken' Breccia (pronounced, BRET-cha). Chunky Conglomerate and Broken Breccia are fraternal twins. In other words, they look somewhat like each other, but are different.

Broken Breccia is made of bits and pieces of angular (not rounded), broken rock and mineral.

Breccia

Activity

Take out the breccia from your kit and look at it closely. Describe what you see. Record your observations and/or pictures.

Activity

Take out the breccia from your kit and look at it closely. Describe what you see. Record your observations and/or pictures.

'Cunning' Chemical's sedimentary rocks: I use the word *cunning* because the chemical sedimentary rocks that we find have been formed under circumstances that are not happening today. How they formed is a bit of a mystery for modern geologists! Chemical sedimentary rocks are made of chemicals that have been changed into hard rocks. Some of the chemicals come from once-living things like sea shells. Some of the chemicals come from the rock and soil that once covered the earth before the flood.

Chemical Sedimentary Rocks

'Limy' Limestone (no fossils). This limestone has no fossil in it, but is still made of the light-colored mineral calcite.

Limestone

Activity

Take out the limestone from your kit and look at it closely. Describe what you see. Record your observations and/or pictures. Limestone is used around the world for building things. Did you know that many of the great pyramids of Egypt are built of limestone?

'Healthy' Halite (pronounced, HAY-light). Halite is both a mineral and a sedimentary rock – neat trick, isn't it? Healthy Halite is one of the most valuable of minerals. Taste it! Yes, it is salt, rock salt. When the flood came on the whole earth, it left huge deposits and caves of salt. Halite is made of the chemicals, sodium and chlorine. By themselves sodium and chlorine are very dangerous. But God specially designed these two chemicals to work together to give us salt! Salt is not only used for flavoring, but also for preservation. Do you remember when Jesus said, "Salt is good...."? (Mark 9:50)

Did you know that the one of the largest salt mines in the world is underneath Kansas? Just how did all that salt get under there?

Halite

Activity

Take out the halite from your kit and look at it closely. Describe what you see. Record your observations.

'Twirling' Travertine. Travertine is a kind of limestone. It is made when water that has some acid in it passes through limestone underneath the ground. When it does this, it takes some of the limestone with it to the surface of the ground and then leaves it on the ground where it turns to rock. It is then called travertine. Travertine is used as a building stone. The Coliseum in Rome is made of travertine.

The Coliseum in Rome was originally made of travertine.

Mammoth Terraces in Yellowstone National park is made of travertine.

68

Activity

Take out the travertine samples from your kit and look at them closely. Describe what you see. Record your observations. Travertine can wear different colored clothing and sometimes its clothing is stained red to orange because of the dark-colored mineral iron.

'Biota' Biochemical's sedimentary rocks: The word *biota* means the animal and plant life of a particular region. Biochemical sedimentary rocks are made of chemicals that have been changed into hard rocks. But these chemicals all come from things that were alive at one time, like ferns and trees and sea shells.

> **'Fizzy' Fossil Limestone.** Limestone is made up of mud rich in lime or the chemical, calcium carbonate or calcite. This limestone is rich in fossil sea shells. When it comes in contact with a strong acid, it fizzes.

Fossil limestone

Activity

Take out the fossil limestone from your kit and look at it closely. Describe what you see. Record your observations.

> **'Creamy' Chalk.** Chalk is made up of very tiny bits and pieces of diatoms. Diatoms were once tiny living sea creatures. So, chalk is a type of limestone. When the flood came on the earth, billions of these tiny sea creatures were buried in masses of very fine sediments. The White Cliffs of Dover, England are huge formations of chalk!

The White Cliffs of Dover

Chalk

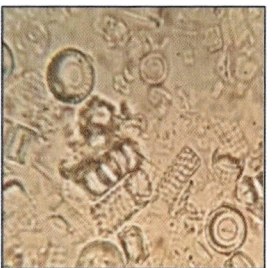

Fossil diatoms magnified 30x, and close-up

Activity

Take out the chalk from your kit and look at it closely. Describe what you see. Record your observations.

'Carby' Coal. Coal is made up of carbon from ferns, leaves and trees that used to be alive. During the flood, they were torn up and buried under a great amount of pressure. And just like big hills of grass and leaves of today, called compost, heat builds up within the pile of dead leaves and grass. During the flood with so many dead plants, a chemical change took place and the pile of leaves, ferns and trees become coal.

Bituminous coal

Activity

Take out the coal from your kit and look at it closely. Describe what you see. Record your observations. The black marks left on your fingers is carbon. Did you know that diamonds are made of pure carbon? No one knows how diamonds formed. But we suspect that they were formed by the same process of heat and pressure, only a lot more of them!

> **'Cryptic' Chert.** Chert can be both a biochemical and a chemical sedimentary rock. Chert is made of the light-colored mineral, quartz. Chert is biochemical when it is made of tiny sea creatures called plankton that had quartz in their skeletons. Chert is chemical when it is simply made of the light-colored mineral, quartz that may have come from quartz crystals. We give it the name *cryptic* because you cannot see the tiny quartz crystals and tiny sea creatures with the naked eye. Cryptic means *hidden.*

Chert

Activity

Take out the sample of chert from your kit and look at it closely. Describe what you see. Record your observations.

Activity

We are going to make a *Sedimentary Sandwich*, to help you understand how this all stacks up!

Gather together:

Two slices of bread

Creamy spread (peanut butter, cream cheese, hummus, etc.)

Sugar, brown or white

Nuts, seeds (alternative could be dried cranberries, raisins, etc.)
Banana slices Plate

1. Spread your creamy spread on one slice of bread. This is your mud layer.
2. Sprinkle a little sugar over the spread. This represents sand.
3. Add a layer of nuts/dried fruit, etc. These individual items represent bones of creatures.
4. Layer banana slices over this. This final layer represents dead vegetation.
5. Place final slice of bread on top. Press down to compress the layers.

You finished sandwich is a representation of the sedimentary layers that occurred at the time of the Flood and after. It is important to note that none of these layers have any dates attached to them. Instead, notice that the bottom layer is the first layer, the second layer is the next, etc. This may seem obvious, but it is an important principle to note. In the field, the bottom layers of sediment are there because they were laid down first. And there are no dates in the field, either! This principle is called the Law of Superposition.

Activity

Since we have concluded the portion of this book that deals with rocks, it would be a good time to make a trip into the field to collect specimens. Using what you have learned, try to identify them.

If it would help you to refer to the rock groupings charts, you can find them all in one place in Appendix B (p. 138).

Section IV – Fossils and Noah's Flood

Part 1: What is a Fossil?

The word fossil comes from a Latin word, *fossilis*, meaning, *dug up*. Many people think that in order to find a fossil we must dig very deep. That is true for some fossils. But many fossils are found on top of the ground or just below the surface of the ground! The picture below is from Eastern Montana where a lot of dinosaur fossils are found. A person can walk across this barren land and find parts of dinosaur fossils all over the ground. Why is that?

Badlands of Eastern Montana

Geologists who do not believe in the Bible or its history look at the picture above and imagine that at one time, millions of years ago, this land was inhabited by dinosaurs. They imagine a tropical forest used to be here with lots of food and many dinosaurs. But then after millions of years of slow erosion (decay and wearing away) by wind, rain and snow, the land of millions of years ago changed into what you see in this picture. This story is taught to kids as a true story. But is it true?

There is another story that is not being told to kids, and that story is about a huge flood that God sent on the earth about 4,500 years ago – not millions of years ago! This flood covered the whole earth and changed it forever. Pictures like the one above tell of a story of land that was torn up by raging flood waters and then deposited all over the world. The picture above tells us that the land was scarred forever because of man's sin against his Creator. That is the reason why you can find fossil bones and other fossils all over the ground. The living things that were caught in the Flood of Noah's day died and then their bodies were torn to pieces and their bones scattered all over in mud. The mud then hardened and became rock. We call these bones *disarticulated* – a big word that means torn apart and scattered.

Look at the next picture. How many bones can you count? Can you tell what kind of creature these bones came from? This is how most fossils are found.

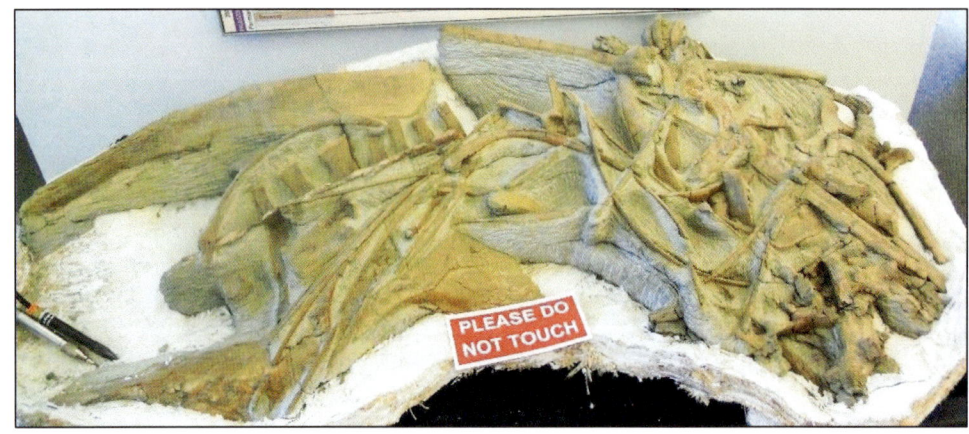

Now look at the following picture. This picture shows fossils from the sea that have been torn up, moved around and buried in mud that became rock.

Which story is true and which is false? This may surprise you, but science cannot answer that question! Science cannot go back into the past and record what happened. There is only one person who can tell us what happened. That person is God. And He did tell us in His Bible, the Book of Genesis. That book does not tell us everything that happened, but it tells us enough to form a picture in our minds. God's word, the Bible, is not man's imagination, but a historical account of what happened 4,500 years ago. Man's word is imagination and his story conflicts with God's story.

A fossil is a record of life that once lived. A fossil is usually hard rock. But many fossils have been found that are still a little soft and have not turned to stone yet. Does it take a long time for something to turn to stone? People used to think that. But that is not true. It takes the right kind of circumstances or chemical environment for something to become a fossil. In other words, something that has just died must be buried quickly in mud for it to be preserved.

What happens to an animal when it dies? Well, other animals come and have lunch, don't they? The body of that animal will soon disappear. Or if the dead animal remains unburied, decay will take over and the animal will eventually rot and disappear. That is why the Genesis Flood explains the fossils we find better than modern science does. The Genesis Flood would have buried billions of living things quickly in mud and kept them from rotting away. Below are pictures of some fossils that must have been buried quickly. Why do we say this? If you have seen a fern die, it curls up and turns brown and then blows away. The fossil ferns in the next picture are flat and fresh-looking.

Fossil fern

The next picture shows fossil clams. Do you notice anything interesting about them? They are closed! When clams normally die, their shells open and in time the two halves of the shell will become *disarticulated* and broken and drift away. But there are
many fossil clams I have found that are complete, as if they were buried suddenly and while they were still alive.

Fossil Clam

The environment for the fossils shown above was one of violent turbulent water and mud that rapidly buried billions of living things in mud we call *sediments*.

So, it does not take a long time for something to become a fossil. It takes the right kind of conditions – quick burial in *sediments* to keep the dead creature from being picked apart by other animals and from being torn apart by bacteria and decay.

Activity
Make a fossil imprint Gather
the following:
Small shallow disposable plastic container (four in. x four in. is a good size.)
Vaseline
Plaster of Paris
An item for imprint - leaf, twig, or shell

Process:
1. Using the Vaseline, coat the leaf (twig or shell) with a thin coat of Vaseline. Lay aside.
2. Mix a small amount of Plaster of Paris. You will need to have a finished product that will be deep enough to accommodate an impression of the imprint item you have chosen. Usually 1 in. deep is deep enough.
3. Pour Plaster of Paris into the plastic container.
4. Gently press imprint item into the Plaster of Paris. Do not submerge it.
5. Let the imprint you have made set up until it is mostly set. This can take as little as five minutes, depending on the consistency of your Plaster of Paris. It usually takes about 15-20 minutes.
6. Carefully remove the imprint item, and allow the Plaster of Paris to continue to dry. It should be completely dry in 24 hours.
7. Remove your imprint from the plastic container.
8. Paint it, if you like!

Section IV – Fossils and Noah's Flood

Part 2: Types of Fossils

Types of Fossils

Remember, fossils are records of things that once lived and are now preserved in rock. Are there different types of fossils? Yes, many different types.

Some fossils are petrified remains

The word *petrified* means *to turn to stone or rock*, from the Greek word for stone or rock, *petros*. How does that happen? Geologists do not know for sure. But we can guess. One of the first things that must happen is for there to be enough water and mud that is rich in the right kinds of minerals. Next, the water and mud that is rich in minerals must be able to enter the dead plant or animal's cells. The cell of a living thing is filled with fluid or liquid. In petrification, the cell's fluid is replaced by mineral-rich water like lime mud. This is why the Bible's explanation of a huge flood makes sense. This event would have provided the necessary water and mud for a dead thing to be preserved.

Once-living things that have been petrified or turned to rock: fish, dinosaur eggs and brittle star.

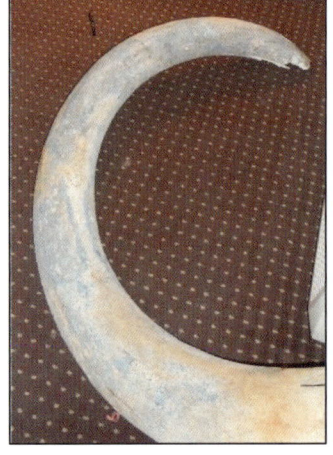
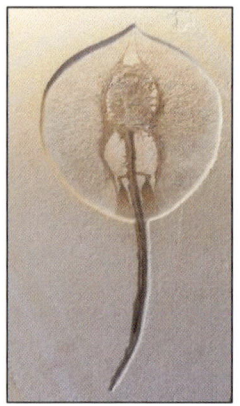

Once-living things that have been petrified or turned to rock: tusk and ray

Did you know that tiny baby dinosaur fossils have been found in fossil dinosaur eggs?

Some fossils are mineralized remains

This is very similar to petrification. But the minerals that have replaced the cells of the dead thing usually contain the minerals quartz or calcite. Petrified wood and fossil sea creatures are very good examples of this.

Cells of wood that have been replaced by quartz (agate) and sea creatures whose cells have been replaced by calcite

Some fossils are casts or molds

A cast is made in soft mud. It shows us where a creature used to be. It is not the creature, but is a trace of where the creature was buried. A mold is very similar. A fossil mold is formed when an animal, plant, or other organism dies and is covered by mud. Its flesh decays and its bones deteriorate due to chemical reactions. The once-living creature decays away, but leaves a cavity with evidence that it once lived.

A mold and cast

Casts and molds are a kind of fossil called an *ichno fossil*. The word *ichno* is a Greek word meaning *track*. We also call it a *trace* fossil. These fossils provide *evidence* that the organism once lived. They are not the actual creature, but things left by the creature that tell us he has been there. These kinds of fossils include tracks, burrows and gastroliths or stomach stones. A stomach stone is a stone swallowed by an animal when it was alive. It was used to help the animal digest his food. Chickens do this today.

A track tells us that a dinosaur or some other animal once walked in soft mud

Fossil worm burrows tell us that worms once wiggled in soft mud

Gastroliths, meaning, *stomach stones*, polished inside of creatures as these stones helped in the digestion of food

Activity

In your kit, you have some fossils that are like the ones in the pictures in your lesson. Take out those fossils and look at them closely. What do you see? Record your observations. Try to identify these fossils according the categories from the lesson: petrified, mineralized, casts, molds, or trace (ichno) fossils.

Section IV – Fossils and Noah's Flood

Part 3: Famous Fossils

In this lesson, you will see some pictures of some amazing fossil finds! Your activity for Lesson 3 will be to study these pictures and discuss them with an adult. Discuss questions like, "What makes these fossils special?" "What do you notice about these fossils?" So be sure to look at them closely!

This fossil is on display at the Royal Tyrrell Museum. What do you notice about its neck and head? This unusual fossil shows that the dinosaur died a
very violent death. The arched-back neck and head is called the *death pose*. It indicates sudden death by suffocation. Could it be that this dinosaur died in the muddy waters of the Genesis Flood?

For many years, it was thought by geologists that the Coelacanth fish had gone extinct along with the dinosaurs. Then in the 1930s a live Coelacanth was caught off the coast of Madagascar – just like its fossil!

This fossil is a trilobite. Do you know why it is called a trilobite? Its body is divided into three parts. Why do you think this is an amazing fossil?

This following fossil is called a *brittle star* and they still live today! They are soft and fragile creatures. Why would a fossil like this be unusual?

These fossil fish have been preserved in a kind of lime mud. These are very common fossils. They are especially found in abundance in southwestern Wyoming. What kind of things do you think make this fossil famous?

What kind of creature is this next one? Geologists think this shrimp is millions of years old, but modern shrimp look no different!

Several years ago, the BBC News reported this amazing find. It is a fossil squid that still had liquid ink in its sack! Here is the question – How is it possible to have something as soft and sloppy as an ink sac fossilized in three dimension, still black, and inside a rock that is 150 million years old?[1]

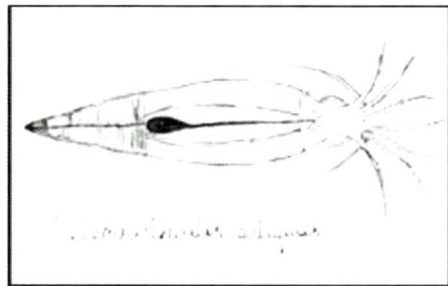

Why might this fossil be famous? It is of an extinct fish giving birth to a baby fish! This fossil is certainly better explained by the Genesis Flood, because for this event to have been preserved, both fish would have had to be buried quickly and while the fish were still alive! What else do you notice about this event?

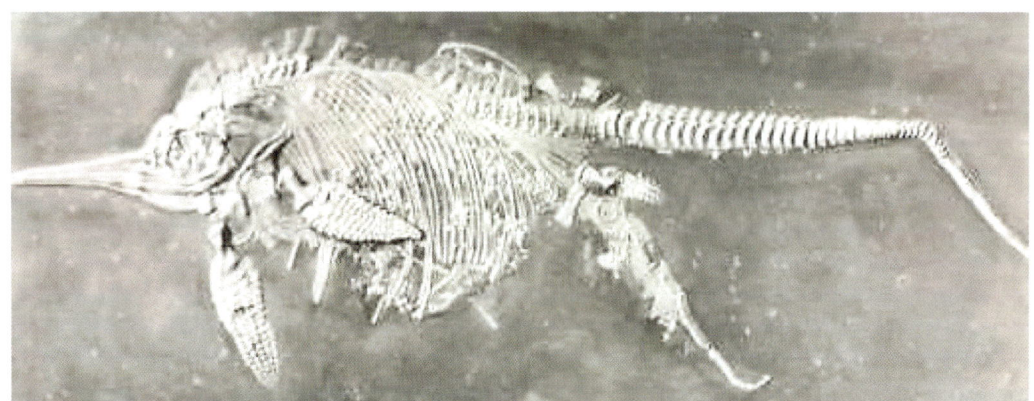

---

[1] Aug. 19, 2009. http://news.bbc.co.uk/2/hi/uk_news/england/wiltshire/8208838.stm.

The following fossil shows a fish in the act of eating another fish. The Genesis Flood can easily explain this. The fish would have to have been alive when buried.

In the 1990s a paleontologist discovered soft tissue in a Tyrannosaurus Rex bone. She thought the bone must be at least 65 million years old. But if the bone was that old, could the tissue have remained soft for that length of time? The tissue and tendon is still soft and elastic. Why is this fossil important?

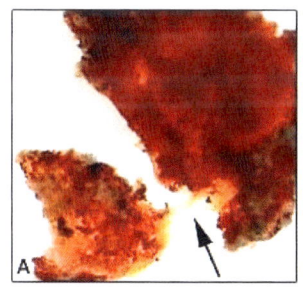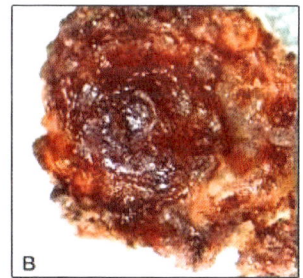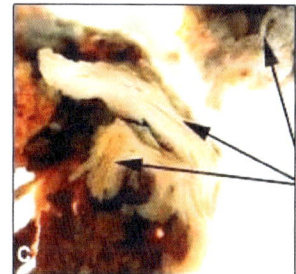

In the early 1900s a rather unusual tooth was discovered in Nebraska. It was sent off to the famous paleontologist, Henry Fairfield Osborn. In 1922 Osborn proclaimed that it was a tooth of a missing human link and named it *Hesperopithecus, ape of the western world* or, affectionately known as *Nebraska Man*. The news spread like wildfire and the picture on the right appeared in a British newspaper, complete with its mate and habitat. Not too long after this, it was discovered that the tooth was that of an extinct pig!

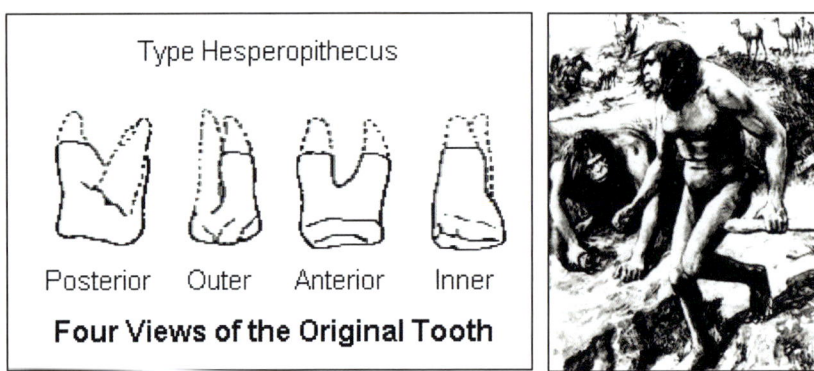

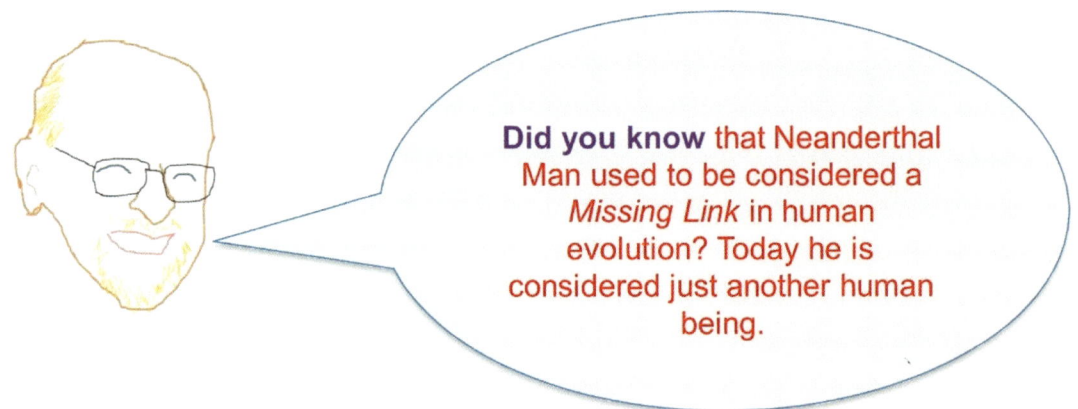

Activity

Study the pictures from this section and discuss them with an adult. Ask questions like: What makes these fossils special? What do you notice about these fossils? Do the same thing with the fossils in your kit.

Activity

Using either your computer or poster board, make a display of famous real or fake fossils. Use magazines, the internet, drawings you create, etc. Be creative! Leave room to add in *Living Fossils*, which we will discuss in the next section.

Section IV – Fossils and Noah's Flood

Part 4: What do Fossils Really Tell Us?

Many geologists teach us that fossils tell a story of evolution. Evolution is the belief that life arose on this earth over several billion years without God and then changed over hundreds of millions of years into different kinds of creatures. This story also teaches that man changed from ape-like creatures into what we are today – again, without God.

Besides being a silly story, it is a dangerous story! It says that the story told in the Bible is false! But wait a minute, if the Bible story of creation is false, then so is what the Bible says about Jesus and why He came. That is dangerous because Jesus is the only way we can be saved from God's judgment. So, it is important that we get this story of creation correct!

What fossils don't tell us!

Let's begin by understanding first that fossils do not show us that life developed over hundreds of millions of years. In order for evolution to be shown scientifically, there would need to be all kinds of fossils found that show us the changes that took place for one creature to become another different creature. THERE ARE NO FOSSILS THAT EVEN COME CLOSE TO SHOWING THAT THIS HAPPENED! In fact, many paleontologists (those who study fossils) are convinced that the fossils do not support what Charles Darwin taught. Charles Darwin was the person who wrote a very famous book that taught that all living things are related to each other and have changed over millions of years from each other. Charles Darwin did not believe in the Bible or the Genesis Flood. Fossils cannot tell us about where the plants and animals lived. Because the Flood was a violent event in which billions of living things were torn up, transported and buried during the Flood, we do not know exactly where they lived. The earth before the Flood was nothing like the earth after Flood. The entire surface of the earth was changed forever.

One very famous fossil was thought at one time to be a missing link between dinosaurs and birds! His name was archaeopteryx (pronounced, *ar-kee-OP-ter-icks*). But today he is now called an extinct bird with some strange characteristics. Below is the fossil of this famous bird.

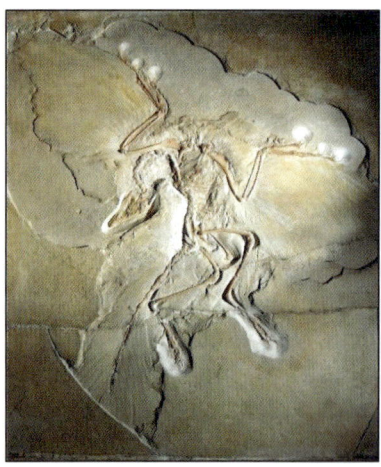

You can see the feathers and wings of archaeopteryx. Do you notice the arched-back neck and head? What did we say this was called? This archaeopteryx must have died a quick and violent death to be so perfectly preserved as a fossil. Such fossils tell us that in the past there was once a big global flood that destroyed all land-dwelling creatures because of man's sin. The fossil record is a record of death and violent destruction, not of evolution!

What fossils do tell us!

Since fossils are remains of living things that once lived, they can tell us a little about what kind of creatures they were and a little about how they lived and what they ate. And fossils tell us that those creatures were specially created just like the Book of Genesis says – kinds reproducing after their kinds. Some of the plants and animals have gone extinct, but many still live. And the plants and animals of the fossil record show that they were wonderfully designed and made by an all-powerful and all-wise Creator.

There are lots of varieties of fossils in the record. But they all belong to their respective kinds. Some of these animals still live today, while others have gone extinct, just like living things do today: ray, fish, fish, bat, turtle and bee.

> **Did you know** that *dinosaur mummies* have been found? They are bones and skin preserved together.

Living fossils

What are living fossils? These are creatures that were once thought to have gone extinct millions of years ago, but then are discovered alive and looking exactly like their fossil! These kinds of fossils are not rare. There are over 500 different living fossils that have been discovered. Here are just a few.

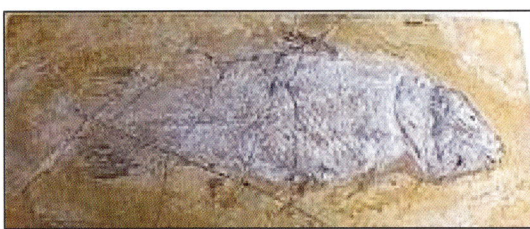

We looked at these pictures in our last section. The fossil on the left is of a coelacanth fish once thought to have gone extinct with the dinosaurs, 65 million years ago. Then in the 1930s live coelacanth fishes were caught off of Madagascar. He looks just like his fossil – no evolutionary change at all!

This gingko fossil is supposed to be 200 million years old. It was once thought to have gone extinct millions of years ago. But then it was discovered alive and unchanged from its fossil

   The Horseshoe Crab – virtually unchanged from its fossil supposed to be millions of years old. Many think this creature is the living survivor of the trilobites who most likely perished in the Genesis Flood.

   The crocodile – unchanged from its fossil which is the same age as the dinosaurs are thought to have been. Could it be that he simply survived the Genesis Flood and the Post Flood Ice Age? Maybe he is just not that old.

The fossil in the center of the picture is of a brittle star. The creature lived at the time of diplodocus, the skeleton on the left. The creature on the right is a living brittle star. Why did he survive and diplodocus did not survive?

Activity

With a parent, do some research and find other living fossils. Have your mom or dad look up Creation Ministries International on the Internet. In the search box, type in the words, *living fossil*. Read the article together. An excellent book on this subject is <u>Living Fossils</u> by Dr. Carl Werner and Carla Axarra. Add the results of your search about *Living Fossils* to the project you began in the last section.

Section IV – Fossils and Noah's Flood

Part 5: Where do We Find Fossils?

Fossils are fun to collect. It is exciting to spend time digging through sediment and then discovering something that was once alive now turned to stone. Once the fossil collecting bug bites you, you will be hooked! So, where do we find fossils?

First of all, we have to talk about the kind of rocks that will most likely contain fossils. Since fossils were formed during the Genesis Flood, then they could be anywhere. But one rock type will be the best type of rock in which to find fossils. Can you guess which rock type that would be? If you said sedimentary rocks, you would be right.

There are four kinds of rocks, but only **one** is the best rock for preserving the remains of once-living things – the sedimentary rocks.

Plutonic rocks were the foundation rocks put together by God before there was any living thing on Earth. These were most likely created during the first couple of days of creation. As we said earlier, these are the granite type rocks or what we call, *basement* rocks.

Most if not all granite was formed before God had created life on Earth.

Volcanic rocks were formed during the Flood and are being formed today. Most living things would have been burned up when coming into contact with the hot lavas. But occasionally lava will cover a tree, for example. Of course, the tree is burned up, but it does leave a mold of itself in the lava.

Lava   Mold of tree

Metamorphic rocks are rocks that have been changed by heat and pressure. Most of the time fossils would have been melted during this time and there would normally not be any traces of fossils left. But occasionally, a fossil is found preserved in metamorphic rocks.

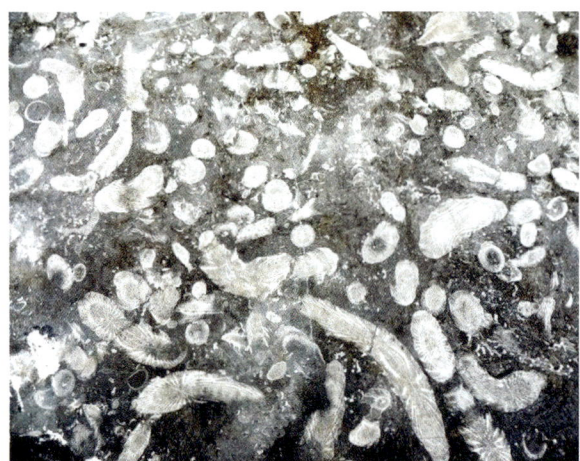

Fossils that have been preserved in metamorphic rocks have shown signs of heat and pressure in bending them out of shape and generally making them unrecognizable.

Sedimentary rocks are the best kind of rock to find fossils in. Their name means *sediment* which is made of lime mud or other kinds of mud. There is no change dealing with heat or pressure. The once-living organisms are simply buried and preserved in rock that was once mud.

Fossil crinoids and brachiopods in limestone

Sedimentary rocks can be found everywhere. I have found them in the Cascade Mountains which are mostly volcanic and metamorphic rocks. But since the earth was covered at one time with lots of water and mud, it should not surprise us to find sedimentary rocks anywhere!

Where to look for fossils?
Geologists have done a great job of mapping rock formations all over the world. I use a book called, *The Roadside Geology of...* and then I pick the state, such as, *The Roadside Geology of Washington*. These books contain a lot about evolution, but they do show you where and what the rocks are in each state along highways. The best place to look for rocks with fossils is in road cut outs, especially along US Highways. Have your mom or dad locate a US Highway with a road cut- out and then pull off and start looking. Be careful though, as there is moving traffic all the time. I have pulled

off the road many times and found buckets full of marine fossils. Marine fossils are fossils of sea creatures. And most of the fossil record is made up of sea creatures!

> **Did you know** that some of the best states to find marine fossils in road cut-outs are Kansas, Indiana, Ohio, Kentucky and Iowa? Notice where they are in relation to the oceans.

Kinds of plants and animals that were fossilized

Fossils represent four kinds of living things. Two of these kinds are invertebrate fossils and vertebrate fossils. Invertebrate fossils were animals that did not have any backbone. These included the trilobites, brachiopods, clams, crinoids, corals, sea urchins, bryozoans and ammonites.

Activity

Take out the bag of invertebrate fossils and look at them carefully. Can you find matching pictures in an encyclopedia or on the Internet?

The second kind of fossils found is vertebrate fossils. These were animals that had backbones. These included mammals, sea mammals, dinosaurs, amphibians, fish, reptiles, birds and lizards. These are not as common as invertebrate fossils. The land creatures were the last to be buried by the Flood and their bodies were probably floating on the sea for some time as other animals made quick lunch of them.

Activity

Take out the bag of vertebrate fossils from your kit and look at them carefully. Can you find matching pictures in an encyclopedia or on the Internet?

The third kind of fossils found is the plant fossils. There are a lot of these in the fossil record. These of course would include trees (wood), bark, coal, ferns, twigs, leaves and pine cones.

Activity

Take out the bag of plant fossils from your kit and look at them carefully. What do you see? Record your observations.

The fourth kind of fossils found is the algae. These are quite abundant in the fossil record and can be found where ever sedimentary rocks are found. Among the sedimentary rocks, the best rocks to find fossils in are limestone and shale.

Activity

Take out the fossil algae from your kit and look at it carefully. What do you see? Next take out the limestone and shale from you kit. Can you find any fossils in these samples? Record your observations.

Activity

Find the fossils display card in Appendix A. Copy this on to heavy cardstock. Arrange and/or glue your fossil samples onto the card, one per box. Be sure to identify the kind of fossil (invertebrate, vertebrate, plant) and its name (algae, brachiopod, petrified wood, etc.). Do not glue down the dinosaur fossil. Save it for the next section.

Section V – Dinosaurs and Noah's Flood

Part 1: What are Dinosaurs?

These beasts have fascinated people of all ages for hundreds of years. I found my first dinosaur bone when I was in 2nd grade and I have not been the same since!

What is a dinosaur? Are they really hundreds of millions of years old? Why aren't they mentioned in the Bible if they were a part of the original creation? There are lots of questions that I know you have, so let's dig in.

The word dinosaur is actually a fairly new word. It was invented in 1842 by this man, Sir Richard Owen. He was an anatomist (one who works with skeletons) and a paleontologist (one who studies fossils.) He believed in God, although it is not clear whether he believed in the Genesis Flood. He disagreed with Charles Darwin about evolution and spoke out against his ideas.

In those days, Latin, Greek, German and French were the scientific languages. And so, Sir Richard Owen chose a Greek word to describe what had been called a dragon for centuries before. The word *dinosaur* means, *terrible lizard*. Although today it is not clear what dinosaurs actually are, scientists no longer believe they are lizards. Many paleontologists have even considered them to be mammals! Most consider them to be some sort of reptile.

At any rate. dinosaurs have not been caught so that we can study them, so we don't know exactly what they are or what their habits are. It is clear, however from historical accounts that people have seen dinosaurs in recent times.

This picture of a dinosaur called Dracorex looks like what we call a dragon doesn't it?

The next picture shows the skull of a Dracorex that was discovered.

Geologists put Dracorex in the class of dinosaurs called *Pachycephalosaur* or *bone-headed lizard*. He would have been related to this Pachycephalosaurus below.

Pachycephalosaurus and Dracorex

The heads of Viking Ships were often carved to look like dragons. Do you think that the pachycepalosaurs might have been the model for these?

What we do know from the fossil record is that dinosaurs were abundant at one time in Earth history. And then for some reason most of them died out. Could that reason have been the Genesis Flood? Changing weather patterns immediately after the Flood could have played a part in the disappearance of dinosaurs after the Flood. Ever since *the Fall* in Genesis 3, death and decay have been a fact of life. Extinction has been a consequence of Adam and Eve's sin. The Genesis Flood provides a great explanation for the extinction of most of the dinosaurs and for the eye witness accounts of dragons or dinosaurs after the Flood. Only those on the ark would have been saved from the Flood. And only those brought safely through the Flood would have produced offspring.

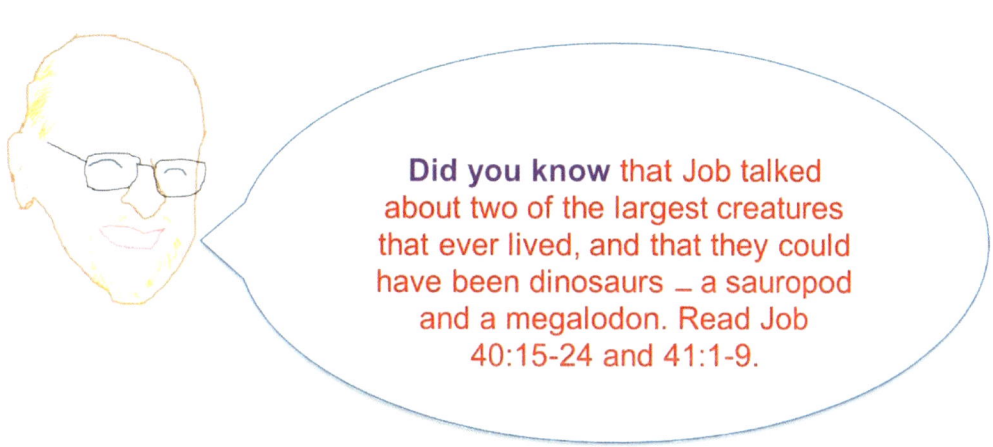

Did you know that Job talked about two of the largest creatures that ever lived, and that they could have been dinosaurs — a sauropod and a megalodon. Read Job 40:15-24 and 41:1-9.

Activity
Take out the dinosaur fossils from your kit and look at them closely. What can you say about the dinosaur fossils in your kit? Record your observations. Add the dinosaur bone to the card from the previous section.

Activity
Read Job 40:15-24 and 41:1-9. Can you draw a representation of what these creatures might have looked like based on the descriptions given here?

Section V – Dinosaurs and Noah's Flood

Part 2: Kinds of Dinosaurs

During the 1700s a man by the name of Carl Linnaeus (1707-1778), a Swedish botanist (one who studies plants), physician, and zoologist (one who studies living animals), came up with a way to organize plants and animals. His system was called, The Linnaeus Classification System. This system uses one of the science languages, Latin to help him organize plants and animals. His system was used to organize LIVING plants and animals. After all, it is hard to classify dead things. You cannot observe their eating or living habits and you cannot observe how they bred.

Carl Linnaeus was a Christian and believed in the Bible. He believed that God had created everything and therefore we could study it and, like Adam, organize God's creation.

During this time a change began to happen in our world. Man began to stop believing in God and His Bible and started to trust his own ideas. Linnaeus' classification system began to be used to organize the fossils and dinosaurs. But because these things were dead and turned to stone, the results were based on man's ideas. And because they are the opinions of men, they have changed time and time again through the years.

So, it is time to go back to Genesis. In Genesis 1 God uses the word, *kind* to describe living things. And then God describes what He means. The kinds were to *reproduce after their kind*. So, a good definition of kind might be that plants and animals be able to reproduce. But since no one has described just how dinosaurs did that, we must guess. Scientists think that they know, but they are just guessing. They don't tell you they are guessing. And so we can be fooled by what they say.

I think there are some things that might allow us to guess as to the kinds of dinosaurs that God had created. Of course, we don't know for sure but we can guess. And I am telling you that I am guessing so you won't be fooled.

When I was a kid, my favorite dinosaur was triceratops (pronounced *tri-SAIR-uh-tops*). Do you have a favorite dinosaur? Triceratops had a shield of bone that covered his head and he had three horns. His skeleton looks like this.

Triceratops

Triceratops' name in the Greek language means, *three horned face*. If you take away the *tri*, then you have *ceratops*, meaning, *horned-face*. There are other dinosaurs that share part of triceratops' name. For example, *protoceratops* (pronounced, *pro-toe-SAIR-uh-tops*. His skeleton looks like this.

Protoceratops

Notice that protoceratops has a bony shield covering his head, but no horns. Yet, he still has the word, *ceratops* in his name. But since we cannot know much about protoceratops, we have to guess. But let's guess based on God's Bible. That would be the safest thing to do because we do not know for sure. Scientists think they know for sure, but they are just guessing. They do not tell you they are just guessing. What do scientists think? The word protoceratops means, *first horned face* or *before the horned face*. Scientists think that the first *ceratops* had no horns at first and then he changed over millions of years to having three horns. But again, they are just guessing. And their guess is based on an idea that is not found in God's Bible. God did not create things to evolve or change into completely different things. He created them to reproduce after their kind.

So, what can we say about this dinosaur that is based on God's Bible? Again, we are guessing. Could protoceratops have been a young triceratops that had not grown horns yet? He could have been. Or perhaps he was a completely different kind. We will probably never know.

Scientists organize these kinds of dinosaurs as *ceratopsians* (pronounced, *ser-uh-TOP-see-un*). This name would put all the *horned-face* dinosaurs into one kind. Here is a picture of the variety of ceratopsians that have been discovered.

Ceratopsian skulls

Since we cannot work with living ceratopsian dinosaurs, we must guess. We can safely guess that these dinosaurs were simply different varieties of the ceratopsian kind. Some of these dinosaurs looked pretty strange.

Now, how would Noah choose a mother and father for the ark? Well, if the ceratopsian dinosaurs were all the same kind, then all Noah would have to do is to make sure he chose a mother and a father. God would take care of the rest.

That is exactly what God did with Noah's family. There were eight people on board the ark. But another way to look at it is that there were four couples or four mothers and four fathers. Together they would have children after the Flood was done. All Noah would have to do to make sure the ceratopsian dinosaurs could have offspring would be to select a mother and father ceratopsian from the horned face dinosaurs.

How many kinds of dinosaurs were there? Well, we don't know for sure, but again, we can guess. Let's see. We have the following kinds of dinosaurs:

1. **The therapods** (pronounced, *THAIR-uh-pods*), whose name means, *beast foot*. So, all the dinosaurs that had feet like this:

 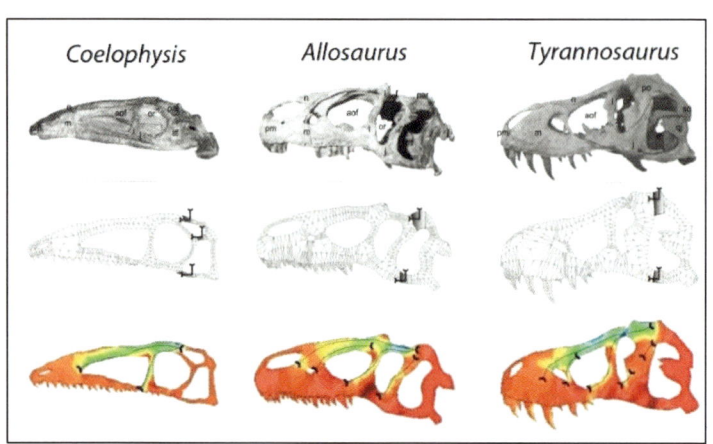

Do you recognize them now? Yes, they are all like Tyrannosaurus Rex.

> **Did you know** that the largest *T. rex* was named Sue, after its discoverer, Sue Hendrickson.

2. **The sauropods** (pronounced, *SORE-uh-pods*), whose name means, *lizard foot*. But that does not make sense, because the sauropods did not have lizard feet! That is silly! The sauropod feet looked like the first picture and lizard feet look like the next picture:

Sauropod foot **Lizard foot**

For the sauropods I like the name, *long necks*. And there were lots of different long-necked dinosaurs.

Sauropods

3. The hadrosaurs, (pronounced, *HA-druh-sores*), whose name means, *thick, bulky lizard*. That does not help very much does it? I like the name, *duckbill* dinosaur. And there were a lot of different duck bill dinosaurs. Now, they don't really have duck bills, but there faces kind of look like duck faces.

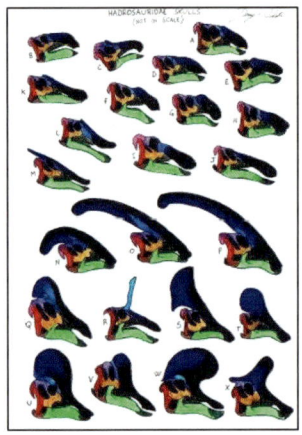 

Hadrosaurs

4. The stegosaurs, (pronounced, *STEG-uh-sores*), whose name means, *roof lizard*. But that does not help much, does it? I like the name, *plated dinosaur*. These were the dinosaurs that had large bony plates along their back.

Stegosaur

5. **The pachycephalosaurs,** (pronounced, *pack-ee-SEPH-uh-lo-sores*), whose name means, *bone-headed lizard* or *the boneheads*. Funny name for a dinosaur, isn't it? But, it describes him well.

Pachycephalosaur

6. **The thyreophora,** (pronounced, *thigh-er-OFF-or-uh*), whose name means, *shield bearers* or, *armor bearer*. This describes the dinosaurs like ankylosaurus, (pronounced, *an-KY-lo-SORE-us*). These were dinosaurs that looked a lot like the armadillo of today.

Ankylosaurus   Armadillo

So here are at least seven kinds of dinosaurs that Noah could have worked with. Most of these dinosaurs were seen by human beings throughout the years since the flood. This means that they must have been on Noah's ark. And that is the subject of our next lesson.

Activity

See how many different kinds of dinosaurs you can find on the internet or in a reference book. Remember to organize them according to kind (theropod, sauropod, etc.) After you organize them according to kind, then see how many different varieties there are among the kinds of dinosaurs. It is a little like organizing your favorite candy. There are kinds of candies, and there are varieties within the kinds of candies. Chocolate might be a kind of candy. How many different varieties of chocolate are there? It would be fun to find out, wouldn't it? Keep track in your notebook of the kinds, and the lists you make.

Section V – Dinosaurs and Noah's Flood

Part 3: Were Dinosaurs on the Ark of Noah?

People make fun of the story of Noah's ark today. But they are blind. They don't think it is a true story. But did you know that Jesus believed in the story? Have your mom or dad read you the story from the gospel of Matthew 24:35-40.

When most people think of Noah's ark, they think it looked like this picture or something like it:

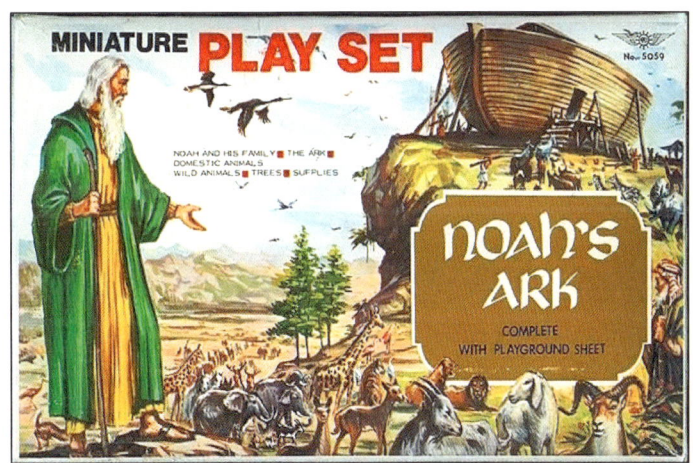

But in the Book of Genesis 6 we are told that Noah was to build an ark (the word ark means, *box*) that was to be 450 feet long, 75 feet wide and 45 feet high with three floors! That is a huge box. Just how big was this ark?

Or another way to look at Noah's ark:

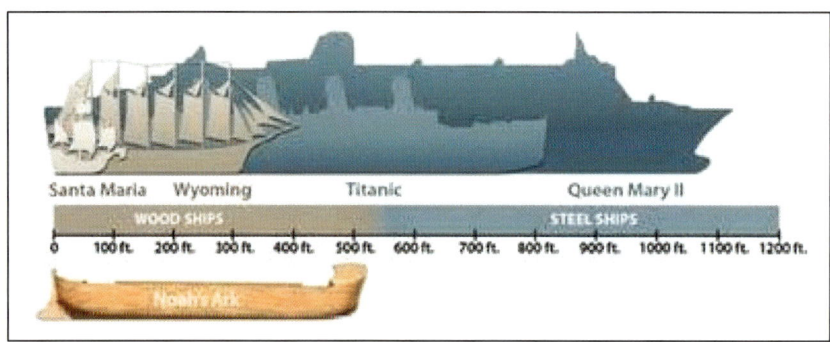

It is clear from the Book of Genesis that the ark was meant to carry lots and lots of animals and food. Do you know how long the flood lasted? The Book of Genesis tells us that it lasted over a year! Noah would have needed this huge ship to carry, house and feed all the animals that were to be on board.

What kind of animals did Noah take onto the ark? Genesis chapter 6 and 7 tells us that Noah took two pairs of every land-dwelling, air- breathing kind of animal. With some of the animals he took seven pairs, those that God told him were clean. I am not sure what those were, but God wanted Noah to take enough of every kind of animal to keep them alive to have offspring after the flood.

What about dinosaurs? Some of the dinosaurs were huge! Well, if I was Noah, I would not take the older dinosaurs, those that were huge and would eat a lot of food. I would take the younger dinosaurs. Dinosaurs were very small when they were born. After they had lived for a while, they grew to be very large. The picture below shows how big a hadrosaur egg was – about the size of an ostrich egg. The next picture shows a 35-foot long hadrosaur. Obviously, he was smaller when he was born! Would you take the small young dinosaur or the big old dinosaur?

Dinosaurs seemed to have grown very much the same way that crocodiles grow! Unless crocodiles die, they will keep growing and growing. But they are small when they are born.

Noah might have taken 7-12 different pairs of parents or 14-24 dinosaurs on board the ark. The ark was large enough to hold all the kinds of animals that lived in Noah's day. The ark was also large enough to hold food and water and Noah's family.

How do we know that Noah took dinosaurs on board the ark? We know this, because people who lived after Noah, reported seeing them! In order to have escaped the flood, these animals would have had to be on the ark. One of those people who saw strange beasts was Job. In the Bible, Job was told by God to think about two beasts He had made. He did this to impress Job with how powerful God was. If God had created creatures that were large enough to scare man, then God must be pretty powerful. Job chapters 40-45 describe two beasts that were very, very big!

The first beast that God described to Job was a creature called, behemoth, (pronounced, *buh-HEE-muth*). The way the creature is described makes him look like a large sauropod dinosaur. The second beast that God described to Job was a creature called, leviathan, (pronounced, *Le-VY-uh-thun*). The way this creature is described makes him look like either a mosasaur, pronounced, *MO- suh-sore* or a megalodon, pronounced, *MEG-uh-lo-don* – huge creatures of the sea.

 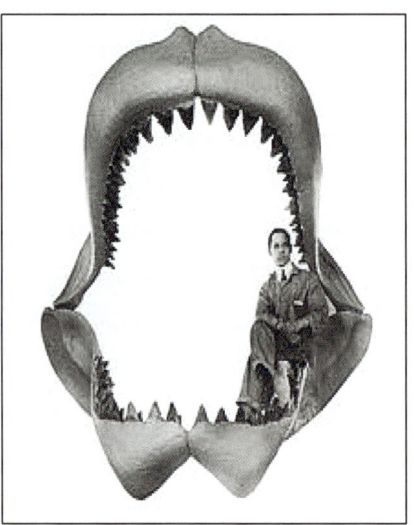

Mosasaur   **Megalodon**

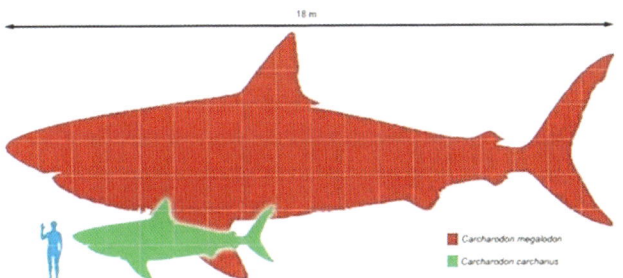

Megalodon was over 54 feet long, and his teeth could be nine inches long!

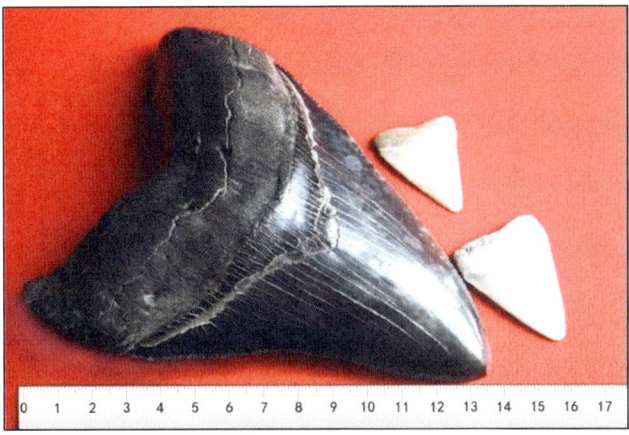

Megalodon tooth

Most everyone is familiar with the story of St. George. During the Middle Ages, it is reported that he bravely killed a *dragon*. Was it a dinosaur that he killed? People knew enough about dragons to report having seen them. I wonder if these could have been dragons descended from those dinosaurs that had been on Noah's ark!

St. George, the dragon slayer; a popular story from Russia

> **Did you know** that there have been hundreds of dinosaur sightings since the flood?

Why don't we see more dinosaurs today?

After the flood was over, in Genesis chapter 9, God told Noah to eat not only plants, but to eat animals too. With that in mind, I wonder if man began to hunt dinosaurs for sport and for food. This could have had a harsh effect on the survival of certain animals. Dinosaurs might have been hunted to extinction!

A second reason might be the dramatic change in weather that came on the earth after the flood. With the large number of volcanoes that were erupting during and after the flood, ash clouds would have kept the warmth of the sun from heating the earth effectively. More moisture from the new warm oceans might have risen into a cooler atmosphere, causing a lot of snow to fall. This would have rapidly brought on an *Ice Age*. We know that millions of mammoths were caught in this weather change all across Siberia and parts of Alaska. Dinosaurs might not have done so well with this extremely uncomfortable weather. Life was also tough for man after the flood. After the Tower of Babel when God scattered man, many looked for caves as their homes. Some tribes of people even died out during this time. People like the Cro-Magnon people and the Neanderthal people might not have survived the harsh weather in the north. Dinosaurs might not have done so well either.

Before the flood, the fossils of dinosaurs and plants seem to indicate that the dinosaurs were used to living in warm tropical climates. After the flood that all changed for hundreds of years until the weather began to warm up again. By then it might have been too late. Dinosaurs and cold weather just don't seem to go together!

Dinosaurs were amazing creatures. And their fossils show that God specially and uniquely created them. Man's sin brought about God's judgment through a worldwide flood. And that flood destroyed a very good world.

Woolly Mammoths and the Ice Age
In addition to dinosaurs, there was another creature that we don't see today that lived after the flood: the woolly mammoth. Although most pictures of the woolly mammoth show him living in cold weather, this is not known for sure. After all, he was an elephant!

And elephants live in the jungle, not in Siberia. In fact, food found in frozen mammoths' stomachs show that it was tropical vegetation. The woolly mammoth might simply have been a variety of elephant that had long hair. What we do know for sure is that millions of them seemed to have frozen to death in the Siberian region as it was changing rapidly after the flood. There is evidence that this is because of an Ice Age that began as a result of worldwide conditions after the flood. Such conditions would have been brought about by massive volcanism from the breaking up of the fountains of the great deep, and the drastic climate change that would have followed.

This event, called the Genesis Flood, is important for us to learn and talk about. It reminds us that each and every person is accountable to God. In other words, God is watching you. When you obey your parents, He is pleased. When you disobey your parents, God is displeased. Some children never learn this lesson and so bring bad things into their lives.

The Flood of Genesis teaches us that every person is responsible to do right things – those things that please God. No wonder modern geologists want to get rid of the Flood story. It was a very important historical event that is meant to show us what happens when people keep doing things that do not please God. And although God promised not to send a flood on the world again, He does want us to see that when we make bad choices, we can bring bad things into our lives and into the lives of others.

Activity

In this activity, you and your parents will read the Genesis Flood story from Genesis chapters 6-9. List the various events that might have anything to do with geology. Record in your notebook the sequence of events and the time involved.

Section VI – The Oceans

Part 1: Ocean Features, Marine Fossils and Corals

Where did the oceans come from? It may surprise you, but this is not a question that can be answered by science. Scientists have offered many ideas on the origin of the oceans, but it remains a mystery among modern scientists.

Did you know that the world's oceans contain enough water to fill a cube with edges over 621 miles in length?

Some facts about the oceans
- Oceans cover around 70% of the earth's surface. That is 140 million square miles!
- The average depth of the oceans is 12,200 feet.
- The deepest point in the oceans is in the Marianna Trench in the Western Pacific at 36,198 feet deep!
- The largest ocean on Earth is the Pacific Ocean, it covers around 30% of the earth's surface.
- The meaning of the word Pacific is *peaceful sea*.
- The Pacific Ocean contains around 25,000 different islands; many more than are found in Earth's other oceans.

- The Pacific Ocean is surrounded by the Pacific Ring of Fire, a large number of active volcanoes.
- The second largest ocean on Earth is the Atlantic Ocean, it covers over 21% of the earth's surface.
- The third largest ocean on Earth is the Indian Ocean, it covers around 14% of the earth's surface.
- During winter the Arctic Ocean is almost completely covered in sea ice.
- While some disagree on whether it is an ocean or just part of larger oceans, the Arctic Ocean includes the area of water that encircles Antarctica.
- The highest mountain in the world is actually not Mt. Everest. It is Mauna Kea, Hawaii which rises 33,474 feet from its base on the ocean floor; only 13,680 feet are above sea level.

In other words, the ocean is a big, big place. Some geologists believe that the oceans came about hundreds of millions of years ago as the earth began to cool from what they say was its original molten state. They believe that as the earth began to cool, the rain stopped hissing away as steam and turned to rain. Over millions of years this falling rain formed the oceans. But just looking at the size of the oceans makes this explanation seem rather silly. Does the Bible have an explanation for the origin of the oceans? It sure does.

The very first chapter of Genesis, the Book of Beginnings, tells us that the Spirit of God hovered over the surface of the water. In other words, God created the earth originally as a ball or sphere of water, not molten magma! The apostle Peter tells us in 2 Peter chapter 3 that the earth was created out of water and by water.

Water is absolutely crucial to our survival and therefore God created it right from the beginning.

Where did the water come from that covered the earth during the Flood? In Genesis chapter 7:11 we read that at the beginning of the Flood the fountains of the great deep burst open. Among other things this would indicate that the waters that now fill the ocean basins came from some other source than the current oceans. The Book of Genesis indicates that the pre-Flood world was totally different than the world of today. Much of the water that now fills our oceans came from deep underground. In Psalm 104:5-9 we read that as the mountains rose at the end of the Flood, the waters rushed off and now fill our present ocean basins.

Features of the Ocean Floor

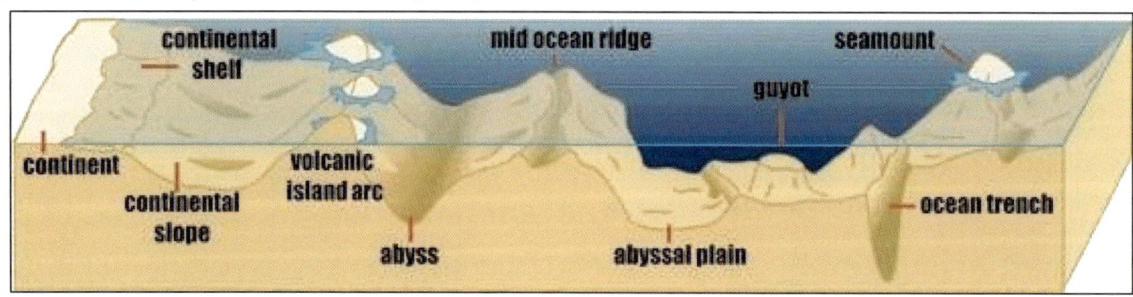

As the Floodwaters raged across the earth, they rapidly eroded the beautiful land that God had originally created. The results were huge scars and craters in the earth. Features like volcanoes and deep trenches called ocean ridges and trenches were formed during this period. Other features such as guyots (pronounced GUY-oats) or flat-topped volcanoes were formed. These flat-topped volcanoes indicate that they were once above water and then sheared as the Floodwaters raced over the earth.

Look at the picture below. It is a map of the ocean floor. Notice the cracks all over the bottom of the ocean floor. Some of these are very deep trenches left over from the breaking up of the fountains of the great deep!

Marine Fossils

Many of our present mountain ranges indicate that they were either formed by huge amounts of watery sediments or that they were below water at one time in the past. Mt. Everest is an example. Marine fossils have been found here at the high elevations!

Mt. Everest; Typical marine fossils

Thousands of well-preserved marine fossils such as trilobites are found high up in the Bolivian Highlands. These trilobites are very easily identified and found in great quantities there.

Examples of trilobites found at the 12,000-foot level in the Bolivian Mountains. Everything right down to the eyes is wonderfully preserved. These trilobites must have been buried rapidly and thoroughly for this kind of preservation to take place.

What other kinds of ocean creatures have been preserved in the fossil record?

Fossil Shark Teeth (Otodus obliquus) – how would you like to have a name like that? Sharks continually shed their teeth, and some Carcharhiniformes (the largest order of sharks) shed approximately 35,000 teeth in a lifetime. In some geological formations, shark's teeth are common fossils. The fossil teeth of Otodus are extremely common.

 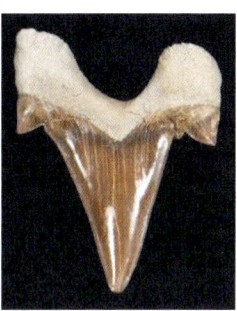

Megalodon - who or what was megalodon? No one really knows for sure because all we have are his teeth. The word megalodon means *big tooth* and they could get as large as 9" long. At any rate, scientists think he is extinct. We will see. People have been surprised on more than one occasion by a creature thought to have gone extinct and then he shows up.

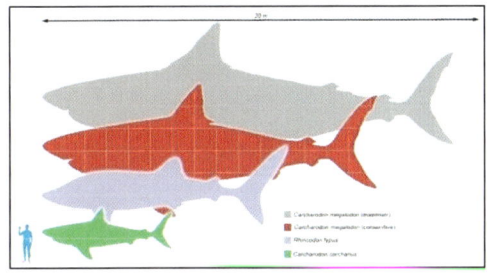 

Fossil Marine Reptile Bones – most likely from a plesiosaur; it is thought to be extinct and so we only have his skeletons to make our guess. He seems to have been quite large though – up to 63 feet long. That is huge.

Fossil Sea Urchins - small, spiny, globular animals which, with their close kin, such as sand dollars, constitute the class Echinoidea of the echinoderm phylum. And guess what? Today's sea urchins look just like their fossil counterparts. According to paleontologists, sea urchins supposedly appeared around 450 million years ago. And yet, they look the same today! Maybe they are just not that old! Can you tell which ones below are 450 million years old and which ones are the modern ones?

Fossil Barnacles - a type of arthropod belonging to the lobster and crab family. Barnacles are encrusters, attaching themselves permanently to a hard surface. Living barnacles have been found in depths as great as 2,000 feet but they are also at home in shallow water. Barnacles are supposed to have originated 450 million years ago, but modern ones look just like their fossils! Are they really that old?

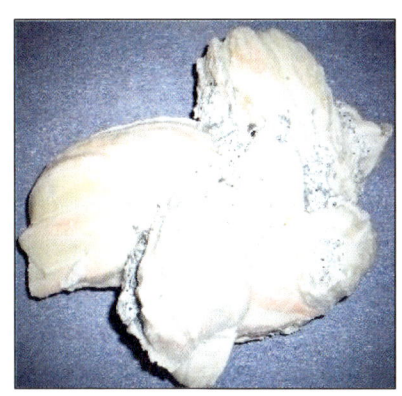 

The Fossil   The Modern

Fossil Worm Tubes – marine invertebrates related to worms. The tube worm does not have many predators, as few creatures live on the sea bottom at such depths. If threatened, the plume may be retracted into the worm's protective tube. The plume provides essential nutrients to bacteria living inside a specialized organ within its body (i.e., trophosome) as part of a symbiotic relationship. They are remarkable in that they have no digestive tract, but the bacteria (which may make up half of a worm's body weight) turn oxygen, hydrogen sulfide, carbon dioxide, etc. into organic molecules on which their host worms feed. They can live up to a few hundred years. Such specialized interesting creatures were designed to do what they do. They did not evolve!

The Fossil Tube Worm – The Modern Tube Worm

Crinoids – also called *sea lilies*, but they are not plants. They are living invertebrates (animals without backbones) related to the starfish. Some have gone extinct, while others are still living. They are highly specialized creatures designed by God to do what they do.

The Fossil Crinoid – The Living Crinoid

Fossil Marine Mammal Bones – The marine mammals include dolphins, whales, porpoises, dugongs, manatees, seals, sea lions and walruses. The bone in your kit is from the dugong, also called a *sea* *cow*. He is a strange looking animal and still lives today in the warmer oceans.

Dugong and his fossil rib bone

Fossil Horn Corals – Rugose is their scientific name. The Rugose is extinct, but the coral family is huge and they all share a couple of things in common. They grow shells in which they live and they attach themselves to the floor of sea beds. Fossil corals are very abundant in the fossil record.

Fossil Horn Coral – Living Coral

Fossil Fish Vertebrae – Fossil fish vertebrae are very common. Fish are some of the most prolific of the sea animals. Beautiful in color and graceful in the way they move about. Such beauties are sure by design and not by evolution. Your fish vertebra in your kit is most likely from the Enchodus, an extinct group of bony fish.

Fossil fish vertebrae

Fossil Fish – Most fish in the fossil record look just like they do today. There are many examples of fossil fish just as there are many examples of living fish today in our oceans. And they are mostly wonderfully preserved, indicating rapid burial.

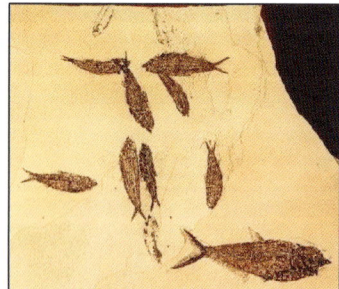

Well-preserved fossils of fish

Corals

Corals are marine invertebrates (animals without back bones) typically living in compact colonies of many identical individual *polyps*. What is a polyp? It is a living animal specially created by God as a unique kind with many variations within its kind. The group includes the important reef builders that inhabit tropical oceans and secrete calcium carbonate to form a hard skeleton. What an amazing creature.

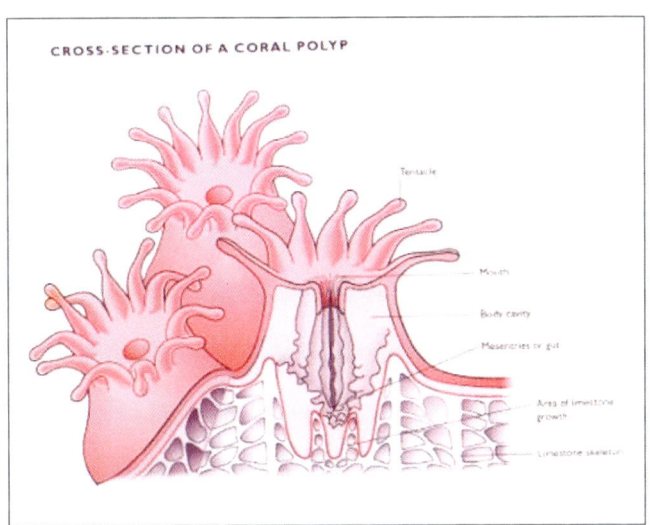

And there are millions of examples of fossil corals, many of which are represented by living corals.

Living Corals

Activity

If there is an aquarium near you, plan a visit to it.

Section VI – The Oceans

Part 2: Ocean Rocks and Sand

What kinds of rocks are abundant on the ocean floor?

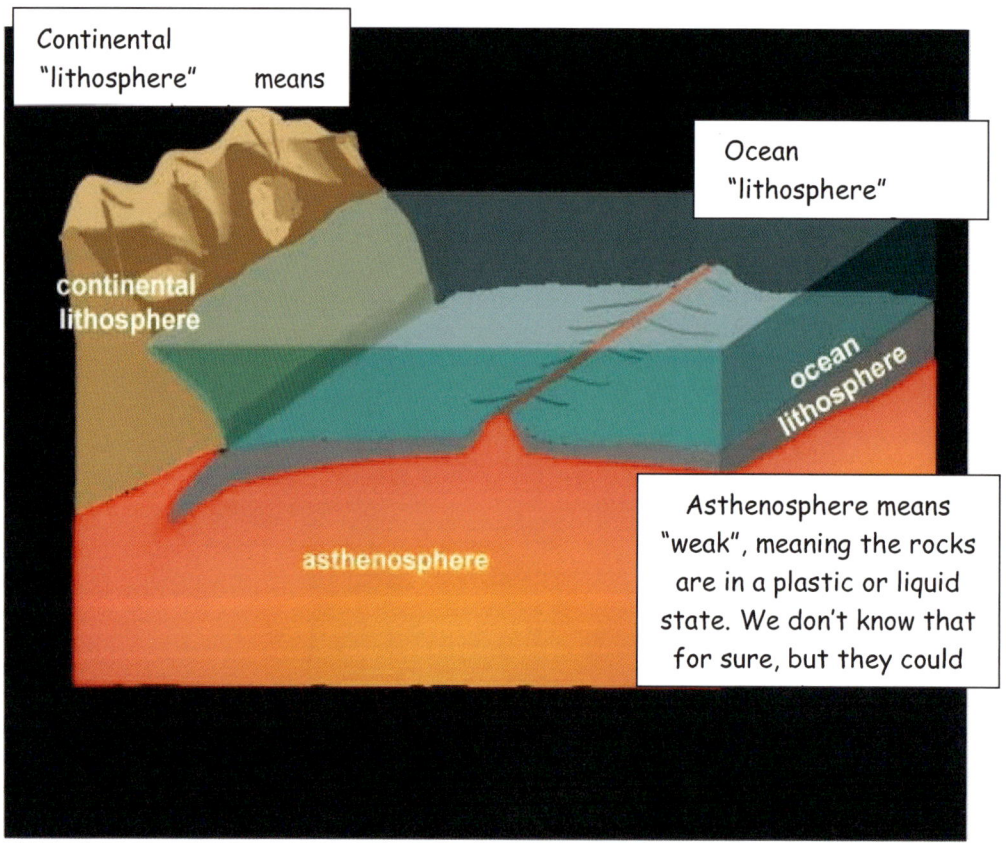

Although geologists have not drilled through the ocean floor, certain rocks might give us a clue as to what lies below the water.

Most people don't think about the rocks of the ocean floor. But they are there and certain kinds are abundant. The rocks of the ocean are what we call *ultramafic* rocks. Mafic stands for magnesium and iron (ferric or ferrous; Latin for iron). So these are rocks high in magnesium and iron. Olivine is an ultramafic mineral as is iron (magnetite), calcium feldspar and pyroxene. These minerals give the

134

rock a dark color. Scientists think these rocks make up the lower crust of our earth. Could be, but we have not been able to scientifically tell this.

Basalt – basalt is a volcanic rock. Do you mean that we have volcanoes on the ocean floor? Yes, we do. And it seems that ever since *the breaking up of the fountains of the great deep* in Genesis 7:11, new basalt lava continues to erupt out of these fissures piling up foot after foot of lava. Basalt is a dark rock containing iron, calcium feldspar, pyroxene and olivine.

Kimberlite – another type of volcanic rock containing a lot of olivine. It is most associated with diamonds!

Diabase – an ultramafic subvolcanic rock. I bet you have never heard of it! It is not a common rock, but it is interesting. This rock has a lot of olivine and pyroxene. The word ultramafic means very high in magnesium (ma) and iron (fic).

Gabbro – another funny name. Gabbro is the name of a town in Italy where the rock type derives its name. It is a coarse- grained rock, dark in color because of the pyroxene and calcium feldspar. It too is identified with the ocean. It is quite common. Just look for a dark-colored rock with big dark mineral crystals in it and you have probably found gabbro.

Peridotite – another ultramafic rock high in olivine. If you will take another look at the name, you will notice the root, *peridot*, which is the gem variety of olivine, a pretty, green mineral.

The Sand of the World's Beaches

Most sand is made up of tiny quartz pebbles. Quartz is an abundant mineral found in granite, which is the most abundant rock of the earth's crust. As the granite is broken up through wave erosion, its hard quartz minerals are tumbled and washed up onto beaches. But there is other kind of sand too.

Black Sand – black sand consists of tiny glassy basalt beads. When the volcanoes erupt their lava into the sea, the hot lava comes into contact with the seawater and it cools and crystallizes immediately. There is one beach in Hawaii where you can see acres of this sand – Punalu'u.

Black sand beach at Punalu'u

Green Sand – Green sand is actually the volcanic mineral peridot. And since Hawaii is a volcanic Island, it would make sense that the tiny grains of olivine should be washed out of the basalt lava as it is churned back and forth by the sea waves. There is one beach in particular that is covered with this green mineral – Papakōlea Beach also known as Green Sand Beach or Mahana Beach.

Magnetic Sand – Magnetic sand is from the mineral magnetite, and it is naturally magnetic! Magnetite is iron oxide, one of three naturally occurring iron oxides. Drag a magnet through this black magnetic sand and see what happens. Remember that the dark lavas contain iron in the form of lodestone or magnetite. As the basalt lava breaks up and erodes, the magnetite is separated and washed up onto the beaches, especially on the coast of Oregon, where the Juan de Fuca volcanic area is active. The darker lavas and minerals like basalt and biotite mica are rich in iron.

Did you know that most of the sand in the world has come from the eroded mountains as they were pushed up at the end of the flood?

Activity

Visit a local beach and see how many different types of sand you can find.

Did you know… that you can email me with your questions about rocks?
northwestexpedition@msn.com
I hope you enjoyed your study!

Appendix A

Vertebrate, Invertebrate and Plant Fossils

Plutonic Rocks	
Granite	Gabbro

Volcanic Lava Rocks	Pyroclastic Volcanic Rocks
Rhyolite	Ash
Basalt	Tuff
	Bombs
	Cinders

Foliated Metamorphic Rocks	Non-foliated Metamorphic Rocks
Gneiss	Quartzite
Schist	Marble

Clastic Sedimentary Rocks	Chemical Sedimentary Rocks	Biochemical Sedimentary Rocks
Shale	Limestone (no fossils)	Fossil Limestone
Siltstone	Halite	Chalk
Sandstone	Travertine	Coal
Conglomerate		Chert
Breccia		

Picture Credits

Section I, Part 2

Hawaii volcano: http://national-parks-virtual fieldtrip.wikispaces.com/Hawai%27i+Volcanoes+National+Park, https://creativecommons.org/licenses/by/2.0/legalcode, 18.

Section I, Part 3
Plutonic rocks: Photo by Patrick Nurre, 20. Volcanic rock: Photo by Vicki Nurre, 20. Andesite lava: Photo by James St. John,
20. http://www.flickr.com/photos/jsjgeology/8455600595/, https://creativecommons.org/licenses/by/2.0/legalcode, CC by 2.0, 20. Dacite lava: http://commons.wvc.edu/rdawes/G101OCL/Basics/BscsTables/dacite.jpg, http://creativecommons.org/licenses/by-sa/3.0/legalcode, CC by-SA 3.0, 20. Rhyolite lava:
http://commons.wvc.edu/rdawes/G101OCL/Basics/BscsTables/rhyolite.jpg, http://creativecommons.org/licenses/by- sa/3.0/legalcode, CC by-SA 3.0, 20. Gneiss:
http://es.wikipedia.org/wiki/Gneis, http://creativecommons.org/licenses/by- sa/3.0/legalcode, CC by-SA 3.0, 20. Schist: Photo by USGS, public domain, found at http://fr.wikipedia.org/wiki/Micaschiste, http://creativecommons.org/licenses/by-sa/3.0/legalcode, CC by-SA 3.0, 20. Quartzite: Photo by USGS, public domain, found at http://es.wikipedia.org/wiki/M%C3%A1rmol, 20. Phyllite: Photo by USGS, public domain, found at http://fr.wikipedia.org/wiki/Phyllite, 20. Limestone: Photo, http://taigaplainsecozone.wikispaces.com/Taiga+Plains, http://creativecommons.org/licenses/by-sa/3.0/legalcode, CC by-SA 3.0, 21. Sandstone: Photo by Vicki Nurre, 21. Breccia. Photo, http://earthscienceteamperiod3.wikispaces.com/RocksGroup6, http://creativecommons.org/licenses/by-sa/3.0/legalcode, CC by-SA 3.0,21. Conglomerate: Photo by Vicki Nurre, 21. Granite, Yosemite National Park: Photo by Patrick Nurre, 21. Mt. Adams: Photo by Vicki Nurre, 22. Agathla Peak: Photo by Patrick Nurre, 22. Glacial Horn: Photo by Patrick Nurre, 23. Beartooth Mountains.: Photo by Patrick Nurre, 24. Cascade Mountains: Photo by Daniel Hershman.
http://creativecommons.org/licenses/by/2.0/legalcode, CC by 2.0, 25. Fossil graveyard: Photo by Patrick Nurre, 26. Fossil graveyard: Photo by Patrick Nurre, 26. Fossil graveyard: Photo by Patrick Nurre, 26.

Section II, Part 1
James Hutton. Painting by Sir Henry Raeburn, public domain, found at
http://en.wikipedia.org/wiki/James_Hutton#mediaviewer/File:Sir_Henry_Raeburn_-_James_Hutton,_1726_-_1797._Geologist_-_Google_Art_Project.jpg, http://creativecommons.org/licenses/by-sa/3.0/legalcode, CC by-SA 3.0, 27. Charles Lyell. Painting by David Octavius Hill, public domain, found at http://pt.wikipedia.org/wiki/Charles_Lyell, 29. Charles Darwin. Photo, public domain, found at http://abhsscience.wikispaces.com/Charles+Darwin, http://creativecommons.org/licenses/by- sa/3.0/legalcode, 29. Dr. Henry Morris: Photo, Fair Use: Publicity photo by the Institute for Creation Research, found at http://creationwiki.org/File:Morris.jpg, 29.

Section III, Part 1
Granite: Photo by Patrick Nurre, 35. Quartz: Photo Thomas Bresson.
http://www.flickr.com/photos/computerhotline/6563772241/, https://creativecommons.org/licenses/by/2.0/legalcode, CC by 2.0, 35. Feldspar: Photo http://www.fierasdelaingenieria.com/los-minerales-mas-peligrosos-del-mundo/, http://creativecommons.org/licenses/by-sa/3.0/legalcode, CC by-SA 3.0, 35. Mica: Photo.
http://creativity103.com/collections/Rock/slides/mica2552.html, http://creativecommons.org/licenses/by-sa/3.0/legalcode, CC by-SA 3.0, 35. Atom model: Image, https://openclipart.org/detail/171335/atom:-carbon-12-by-xsapien171335, 36. Atom model: Image by Vicki. Nurre, 36. Earth's crust: Image by Vicki Nurre, 37. Model of Earth's interior. http://abeach5.wikispaces.com/Earth, http://creativecommons.org/licenses/by-sa/3.0/legalcode, CC by-SA 3.0, 38.

Section III, Part 2
Quartz: Photo by Thomas Bresson. http://www.flickr.com/photos/computerhotline/6563772241/, https://creativecommons.org/licenses/by/2.0/legalcode, CC by 2.0, 40. Granite: Photo by Patrick Nurre, 42. Quartz: Photo by Heidi Ann Noggle, from the author's collection, 42; Jasper: Photo by Heidi Ann Noggle from the author's

collection, 42. Potassium Feldspar: http://allencentre.wikispaces.com/A+Grain+of+Sand CCA-S, 42. Sodium Feldspar: http://geology1a- 1.wikispaces.com/Notes,+3+pictures+and+captions+from+pgs.+87-95, 42. Calcium feldspar: Photo by Heidi Ann Noggle from the author's collection, 42. Biotite mica: Photo by Heidi Ann Noggle, 42; Muscovite mica http://grandjunctiongeology.wikispaces.com/Mica+Mines, CCA-S, 42. Olivine: Photo by Vicki Nurre, 42. Pyroxene: Photo by Vicki Nurre, 42. Amphibole: Photo by Vicki Nurre, 42. Magnetite: Photo by Vicki Nurre, 42 Calcite: Photo by Heidi Ann Noggle, from author's collection, 42.

Section III, Part 3
Plutonic Rock: Photo by Patrick Nurre, 45. Plutonic Rock: Photo by Patrick Nurre, 45. Plutonic Rock. Photo by Patrick Nurre, 45. Granite. Photo by Patrick Nurre, 45. Granite. Photo by Patrick Nurre, 45. Granite carving. Photo by Vicki Nurre, 45. Gabbro. Photo by Patrick Nurre, 46. Gabbro. Photo by Patrick Nurre, 46. Gabbro. Photo by Patrick Nurre, 46.

Section III, Part 4
Mauna Loa. http://www.travelthewholeworld.org/2013/12/top-10-stunning-volcanoes-around-the-world.html, http://creativecommons.org/licenses/by-sa/3.0/legalcode, CC by-SA 3.0, 47. Mt. Shasta. Photo by Vicki Nurre, 48. Crater Lake. Photo by Zainubrazvi.
http://en.wikipedia.org/wiki/Crater_lake#mediaviewer/File:Crater_lake_oregon.jpg, http://creativecommons.org/licenses/by-sa/3.0/legalcode, CC by-SA 3.0, 48. Rhyolite. Photo by Vicki S Nurre, 49. Rhyolite. Photo by Vicki Nurre, 49. Basalt. Photo by Patrick Nurre, 49. Basalt. Photo by Patrick Nurre, 50. Columbia basalts. Photo by Patrick Nurre, 51.

Section III, Part 5
Volcanic ash. Photo by Arni Frioriksson.
http://en.wikipedia.org/wiki/Volcanic_ash#mediaviewer/File:Eyjafjallajokull-April- 17.JPG, http://creativecommons.org/licenses/by-sa/3.0/legalcode, CC by-SA 3.0, 53. Volcanic Ash: Photo by R. Clucas, public domain, found at http://en.wikipedia.org/wiki/Volcanic_ash#mediaviewer/File:MountRedoubtEruption.jpg, 53. Ash. Photo by Patrick Nurre, 53. Volcanic ash magnified: public domain, found at
http://volcanoes.usgs.gov/images/pglossary/ash.php, 53. Volcanic layered ash. Photo by Vicki Nurre, 54. Tuff. Photo by Patrick Nurre, 54. Tuff. Photo by Patrick Nurre, 54. Tuff. Photo by Patrick Nurre, 54. Volcanic bomb. Photo by Patrick Nurre, 54. Cinder. Photo by Vicki Nurre, 55. Cinder cone. Photo by Vicki Nurre, 56.

Section III, Part 6
Quartz. Photo by Vicki Nurre, 57. Gneiss. Photo by Patrick Nurre, 57. Schist. Photo by Patrick Nurre, 59. Marble. Photo by Patrick Nurre, 59. Gneiss. Photo by Patrick Nurre, 59. Gneiss. Photo by Patrick Nurre, 59. Gneiss. Photo by Patrick Nurre, 59. Schist. Photo by Patrick Nurre, 60. Schist. Photo by Patrick Nurre, 60. Schist. Photo by Patrick Nurre, 60. Quartzite. Photo by Patrick Nurre, 60. Quartzite. Photo by Patrick Nurre, 60. Marble. Photo by Patrick Nurre, 61.

Section III, Part 7
Shale. Photo by Patrick Nurre, 63. Sandstone. Photo by Patrick Nurre, 63. Conglomerate. Photo by Vicki Nurre, 63. Shale.
Photo by Patrick Nurre, 64. Shale. Photo by Patrick Nurre, 64. Siltstone. Photo by Patrick Nurre.64. Sandstone. Photo by Patrick Nurre, 65. Quartz crystals: Found at https://allencentre.wikispaces.com/A+Grain+of+Sand, CC by-SA 2.5, http://creativecommons.org/licenses/by-sa/2.5/legalcode, 65. Conglomerate. Photo by Patrick Nurre, 65. Conglomerate: Photo by Heidi Ann Noggle, from the author's collection, 65. Breccia. Photo by Patrick Nurre, 66. Limestone. Photo by Patrick Nurre, 66. Travertine. Photo by Patrick Nurre, 66. Halite. Photo by Patrick Nurre, 66. Limestone. Photo by Patrick Nurre, 67. Limestone. Photo by Patrick Nurre, 67. Halite. Photo by Didier Descouens. http://en.wikipedia.org/wiki/Halite#mediaviewer/File:Selpologne.jpg, http://creativecommons.org/licenses/by-sa/3.0/legalcode, CC by-SA 3.0, 68. Halite. Photo by James St. John.
http://en.wikipedia.org/wiki/Halite#mediaviewer/File:Rock_salt_Mississippi_Potash_East_Mine,_New_Mexico.jpg, http://creativecommons.org/licenses/by/2.0/legalcode, CC by-2.0, 68. Coliseum. Photo by Vicki Nurre, 68. Travertine. Photo by Patrick Nurre, 68. Travertine terrace. Photo by Vicki Nurre, 68. Fossil limestone. Photo by Patrick Nurre, 69. White Cliffs of Dover.
http://en.wikipedia.org/wiki/White_Cliffs_of_Dover#mediaviewer/File:White_cliffs_of_dover_09_2004.jpg, http://creativecommons.org/licenses/by-sa/2.0/legalcode, CC by-SA 2.0, 70. Chalk. Photo by Patrick Nurre, 70.

Diatom: Photo courtesy of Zach Kiser, 70. Bituminous coal. Photo by Patrick Nurre, 70. Chert. Photo by Patrick Nurre, 71. Chert. Photo by Patrick Nurre, 71. Chert. Photo by Patrick Nurre, 71.

Section IV, Part 1
Montana Badlands: http://liceomonjardingeo.wikispaces.com/im%C3%A1genes+tema+3, http://creativecommons.org/licenses/by-sa/3.0/legalcode, CC by-SA 3.0, 73. Disarticulated fossils: Photo by Patrick Nurre,
74. Fossil graveyard: Photo by Patrick Nurre, 75. Fossil fern: Photo by Patrick Nurre, 76. Fossil clam: Photo by Patrick Nurre, 77.

Section IV, Part 2
Fossil fish: Photo by Patrick Nurre, 79, Fossil hadrosaur eggs: Photo by Patrick Nurre, 79. Fossil Brittle stars: Photo by
Patrick Nurre, 79. Fossil tusk: Photo by Patrick Nurre, 80. Fossil ray: Photo by Patrick Nurre, 80. Petrified log: Photo by
Vicki Nurre, 80. Nautilus: Photo courtesy of Tim and Candey, earths. Ancient.treasures, 80. Trilobite: Photo courtesy of Oscar Sanchez, Bolivian Fossils, 81. Dinosaur track: Photo by Patrick Nurre, 81. Fossil worm burrows: Photo by Patrick Nurre, 82. Gastroliths: Photo by Patrick Nurre, 82.

Section IV, Part 3
Death pose: Photo by Sebastian Bergman. https://creativecommons.org/licenses/by-sa/2.0/legalcode, CC by-SA 2.0, 83. Coelecanth fossil: http://earthscienceinmaine.wikispaces.com/11.1+Fossils, http://creativecommons.org/licenses/by- sa/3.0/legalcode, CC by-SA 3.0, 84. Coelecanthe.: http://apenvirotuttle.wikispaces.com/KatieCoelacanth, http://creativecommons.org/licenses/by-sa/3.0/legalcode, CC by-SA 3.0, 84. Trilobite: Photo by Vicki Nurre, 84. Brittle Star fossil.: Photo by Mark A Wilson, public domain, found at http://fr.wikipedia.org/wiki/Ophiuroidea#mediaviewer/File:AsteriacitesUtah.jpg, 84. Fossil fish: Photo courtesy of Tim and Candey, earths.ancient.gifts, 85. Fossil shrimp: Photo by Vicki Nurre, 69. Squid: Photo found at http://news.bbc.co.uk/2/hi/uk_news/england/wiltshire/8208838.stm, August 19, 2009, 70. Fossil Ichthyosaur: Photo, found at http://sixdays.org/Fossils-Confirm-the-Biblical-Creation-and-the-Genesis-Flood, CC by-NC-ND 3.0, http://creativecommons.org/licenses/by-nc-nd/3.0/us/, 86.70. Fossil fish swallowing a fish: Photo, public domain, found at
http://www.nps.gov/media/photo/gallery.htm?showrawlisting=false&id=F17B1C64%2D155D%2D451F%2D676534 1D9B8 E553F&tagid=0&maxrows=20&startrow=21, 87. Soft tissue: Found at http://www.nbcnews.com/id/7285683/#.VJ9issCAA, March 24, 2005, 87. Tooth drawing: Found at http://www.bradburyac.mistral.co.uk/tenness5.html, 87. Nebraska Man. Drawing: Public domain, originally published in London Illustrated news, 1922, found at http://www.bradburyac.mistral.co.uk/tenness5.html, 87.

Section IV, Part 4
Archaeopteryx: Photo by H. Raab, http://en.wikipedia.org/wiki/Archaeopteryx#mediaviewer/File:Archaeopteryx_lithographica_(Berlin_specimen).jpg, http://creativecommons.org/licenses/by-sa/3.0/legalcode, 90. Fossil: Photo public domain, found at http://www.nps.gov/media/photo/gallery.htm?id=F17B1C64-155D-451F-6765341D9B8E553F, 91. Fossil fish: Photo public domain, found at http://www.nps.gov/media/photo/gallery.htm?id=F17B1C64-155D-451F-6765341D9B8E553F, 91. Fossil fish: Photo public domain, found at http://www.nps.gov/media/photo/gallery.htm?id=F17B1C64-155D-451F-6765341D9B8E553F, 91. Fossil: Photo public domain, found at http://www.nps.gov/media/photo/gallery.htm?id=F17B1C64-155D-451F-6765341D9B8E553F, 91. Fossil turtle: Photo public domain, found at http://www.nps.gov/media/photo/gallery.htm?id=F17B1C64-155D-451F-6765341D9B8E553F, 91.
Fossil bee: Photo public domain, found at http://www.nps.gov/media/photo/gallery.htm?id=F17B1C64-155D-451F-6765341D9B8E553F, 91. Coelccanth fossil: http://earthscienceinmaine.wikispaces.com/11.1+Fossils, http://creativecommons.org/licenses/by-sa/3.0/legalcode, CC by-SA 3.0, 92. Coelecanthe: http://apenvirotuttle.wikispaces.com/KatieCoelacanth, http://creativecommons.org/licenses/by-sa/3.0/legalcode, CC by-SA 3.0, 92. Gingko fossil: Photo, http://en.wikipedia.org/wiki/Ginkgo#mediaviewer/File:Ginkgo_biloba_MacAbee_BC.jpg, http://creativecommons.org/licenses/by-sa/3.0/legalcode, CC by-SA 3.0, 92. Gingko leaves: Photo by James Field,

found at http://en.wikipedia.org/wiki/Ginkgo_biloba#mediaviewer/File:Ginkgo_Biloba_Leaves_-_Black_Background.jpg, http://creativecommons.org/licenses/by-sa/3.0/legalcode, CC by-SA 3.0, 92. Horseshoe Crab fossil: Photo by David Goehring. http://commons.wikimedia.org/wiki/File:Horseshoe_Crab_Ancestor.jpg, http://creativecommons.org/licenses/by/2.0/legalcode, CC by-2.0, 93. Horseshoe Crab: Photo by Amanda from Chicago. http://commons.wikimedia.org/wiki/File:Horseshoe_Crab.jpg, http://creativecommons.org/licenses/by/2.0/legalcode, CC by-2.0, 93. Crocodile fossil: Photo. http://fossil.wikia.com/wiki/File:Limusaurus_and_a_small_crocodile_fossils.jpg, http://creativecommons.org/licenses/by-sa/3.0/legalcode, CC by-SA 3.0, 93. Crocodile: Photo by Tomas Costelazo. http://commons.wikimedia.org/wiki/File:Crocodylus_acutus_mexico_02-edit1.jpg, http://creativecommons.org/licenses/by- sa/3.0/legalcode, CC by-SA 3.0, 93. Diplodicus: Photo by Scott Robert Anselmo. http://en.wikipedia.org/wiki/Diplodocus#mediaviewer/File:CM_Diplodocus.jpg, http://creativecommons.org/licenses/by- sa/3.0/legalcode, CC by-SA 3.0, 93. Brittle Star fossil: Photo by Mark A Wilson, public domain, found at http://fr.wikipedia.org/wiki/Ophiuroidea#mediaviewer/File:AsteriacitesUtah.jpg, 93. Brittle Star: Photo by Neil. http://en.wikipedia.org/wiki/File:Greenbrittlestar.jpg, http://creativecommons.org/licenses/by-sa/3.0/legalcode, CC by-SA 3.0, 93.

Section IV, Part 5
Granite: Photo by Patrick Nurre, 95. Fire from lava flow: Photo USGS, found at http://www.brucesussman.com/earthquakes- volcanoes/ring-of-fire-oregon-quakes-hawaii-volcano-eruption/, 96. Fossil tree cast: Photo USGS by C. Heliker, found at http://hvo.wr.usgs.gov/archive/spotlight_images/IMG_6165_20050722.html, 96. Fossils in marble: Photo by Andrew Curtis, found at http://www.geograph.org.uk/photo/1273619, CC by-SA 2.0, http://creativecommons.org/licenses/by- sa/2.0/legalcode, 96. Fossil crinoids: Photo by Patrick Nurre, 97. Fossil brachiopods: Photo by Patrick Nurre, 97.

Section V, Part 1
Dinosaur: Photo by Vicki Nurre, 100. Photo of Richard Owen: Public domain, found at http://es.wikipedia.org/wiki/Richard_Owen#mediaviewer/File:Richard_Owen.JPG, 101. Dracorex: Drawing. http://de.drachen.wikia.com/wiki/Dracorex, http://creativecommons.org/licenses/by-sa/3.0/legalcode, CC by-SA 3.0, 101. Dracorex skull: http://en.wikipedia.org/wiki/Dracorex#mediaviewer/File:The_Childrens_Museum_of_Indianapolis_-_Dracorex_actual_skull.jpg, http://creativecommons.org/licenses/by-sa/3.0/legalcode, CC by-SA 3.0, 102. Pachycephalosaurus: Photo. http://en.wikipedia.org/wiki/Pachycephalosaurus#mediaviewer/File:Pachycephalosaurus_in_Japan.jpg, http://creativecommons.org/licenses/by/2.0/legalcode, CC by-2.0, 102. Pachycephalosaurus: Drawing by Jordan Mallon. http://en.wikipedia.org/wiki/Pachycephalosaurus#mediaviewer/File:Pachycephalosauria_jmallon.jpg, http://creativecommons.org/licenses/by-sa/2.5/legalcodem CC by-SA 2.5, 102. Viking ship prow: http://www.flickr.com/photos/shirokazan/5312752960/, https://creativecommons.org/licenses/by/2.0/legalcode, CC by 2.0, 102. Viking ship: Drawing, http://sandyvikings.wikispaces.com/Jack+and+Anthony's+Viking+Project, http://creativecommons.org/licenses/by-sa/3.0/legalcode, CC by-SA 3.0, 102.

Section V, Part 2
Linnaeus: Public domain, http://commons.wikimedia.org/wiki/File:LinnaeusWeddingPortrait.jpg, CC by-SA 3.0, 104. Triceratops: http://en.wikipedia.org/wiki/Triceratops#mediaviewer/File:LA-Triceratops_mount-1.jpg, http://creativecommons.org/licenses/by-sa/3.0/legalcode, CC by-SA 3.0, 105. Protoceratops: Photo by Karen. http://en.wikipedia.org/wiki/Protoceratops#mediaviewer/File:Carnegie_Protoceratops_andrewsi.jpg, https://creativecommons.org/licenses/by/2.0/legalcode, CC by 2.0, 106. Ceratopsian skulls: http://commons.wikimedia.org/wiki/File:Ceratopsian_skulls.jpg, http://creativecommons.org/licenses/by-sa/3.0/, CC by-SA 3.0, 107. Therapod foot: Photo by Vicki Nurre, 108. Skulls: Rayfield, E. J. 2005. Aspects of comparative cranial mechanics in the theropod dinosaurs *Coelophysis*, *Allosaurus* and *Tyrannosaurus*. *Zoological Journal of the Linnean Society*, 144, 309–316, found at http://ichthyosaurs.wordpress.com/category/biomechanics/, http://creativecommons.org/licenses/by- sa/3.0/legalcode, CC by-SA 3.0, 108. Sauropod foot: Photo by Vicki Nurre, 109. Lizard foot: Photo by Donna Sutton, http://www.flickr.com/photos/77043400@N00/224131637/,

https://creativecommons.org/licenses/by-nd/2.0/, CC by-ND 2.0, 109. Sauropods: Photo by Vicki Nurre, 109. Hadrosaur heads: Drawing by Danny Cicchetti.
http://en.wikipedia.org/wiki/Hadrosaurid#mediaviewer/File:Hadrosauridae_skull_comparison_(not_in_scale).jpg,
http://creativecommons.org/licenses/by-sa/3.0/legalcode, CC by-SA 3.0, 109. Hadrosaur: Photo by Vicki Nurre, 109. Stegosaur: Photo by Eva K, found at
http://en.wikipedia.org/wiki/Stegosaurus#mediaviewer/File:Stegosaurus_Senckenberg.jpg,
http://creativecommons.org/licenses/by-sa/2.5/legalcode, CC by-SA 2.5, 110. Pachycephalosaurs skull:
http://commons.wikimedia.org/wiki/File:Pachycephalosaurus_skull.JPG, http://creativecommons.org/licenses/by-sa/3.0/legalcode, CC by-SA 3.0, 110. Pachycephalosaurus: Drawing by Jordan Mallon.
http://en.wikipedia.org/wiki/Pachycephalosaurus#mediaviewer/File:Pachycephalosauria_jmallon.jpg,
http://creativecommons.org/licenses/by-sa/2.5/legalcodem CC by-SA 2.5, 110. Ankylosaurus:
http://commons.wikimedia.org/wiki/File:FukuiDinosaurMuseum02.JPG,
http://creativecommons.org/licenses/by- sa/3.0/deed.fr, CC by-SA 3.0, 111. Armadillo:
http://commons.wikimedia.org/wiki/File:Armadillo2.jpg, http://creativecommons.org/licenses/by-sa/3.0/,
CC by-SA 3.0, 111.

Section V, Part 3
Noah's Ark toy: Photo by Tom, http://www.flickr.com/photos/tom1231/3563339788/,
https://creativecommons.org/licenses/by/2.0/, CC by-2.0, 112. Comparison of ship sizes:
http://www.queenmarycruises.net/rms-queen-mary-2-ship/, 112. Hadrosaur egg: Photo by Vicki Nurre, 113.
Hadrosaur: http://commons.wikimedia.org/wiki/File:Shantungosaurus-v4.jpg,
http://creativecommons.org/licenses/by-sa/3.0/, CC by- SA 3.0, 113. Baby Crocodile: Photo by Tim Donovan,
http://www.flickr.com/photos/myfwcmedia/6962434859/, https://creativecommons.org/licenses/by-nd/2.0/CC by-ND 2.0, 114. Crocodile: http://commons.wikimedia.org/wiki/File:Crocodylus_acutus_mexico_02-edit1.jpg,
http://creativecommons.org/licenses/by- sa/3.0/, CC by SA 3.0, 114. Mosasaur:
http://es.wikipedia.org/wiki/Tylosaurus,
http://es.wikipedia.org/wiki/Wikipedia:Texto_de_la_Licencia_Creative_Commons_Atribuci%C3%B3n-CompartirIgual_3.0_Unported, CC by-SA 3.0, 115. Megalodon jaw: Public domain, found at
http://en.wikipedia.org/wiki/Megalodon#mediaviewer/File:Carcharodon_megalodon.jpg,
http://es.wikipedia.org/wiki/Wikipedia:Texto_de_la_Licencia_Creative_Commons_Atribuci%C3%B3n-CompartirIgual_3.0_Unported, CC by-SA 3.0, 115. Megalodon comparison:
http://en.wikipedia.org/wiki/Megalodon#mediaviewer/File:Megalodon_scale.svg,
http://en.wikipedia.org/wiki/Wikipedia:Text_of_Creative_Commons_Attribution-ShareAlike_3.0_Unported_License, CC by-SA
3.0, 115. Megalodon tooth: Photo.
http://en.wikipedia.org/wiki/Megalodon#mediaviewer/File:Megalodon_tooth_with_great_white_sharks_teeth-3-2.jpg,
http://creativecommons.org/licenses/by-sa/3.0/legalcode, CC by-SA 3.0, 115. St. George and the dragon: Public domain, found at
http://fr.wikipedia.org/wiki/Saint_Georges_et_le_Dragon_(Rapha%C3%ABl,_mus%C3%A9e_du_Louvre),
http://es.wikipedia.org/wiki/Wikipedia:Texto_de_la_Licencia_Creative_Commons_Atribuci%C3%B3n-CompartirIgual_3.0_Unported, CC by-SA 3.0, 116. St. George and the Dragon: Photo by Klearchos Kapoutsis.
http://www.flickr.com/photos/klearchos/3824329661/, CC by-2.0, 116. Mastodons: Picture by Charles R. Knight, "Restoration of a Herd," Public domain, found at
http://en.wikipedia.org/wiki/Mastodon#mediaviewer/File:Knight_Mastodon.jpg, 118.

Section VI, Part 1
Ocean floor: http://mrbrown.wikidot.com/oceans, http://creativecommons.org/licenses/by-sa/3.0/legalcode, CC by-SA 3.0, 122. World map: http://saulscience.wikispaces.com/Geology, http://creativecommons.org/licenses/by-sa/3.0/legalcode, CC by-SA 3.0, 123. Mt. Everest: http://environmentalsciencehh2.wikispaces.com/High+Mountain,
http://creativecommons.org/licenses/by-sa/3.0/legalcode, CC by-SA 3.0, 123. Marine fossils: Photo by Patrick Nurre, 123. Bolivian highlands: http://es.wikipedia.org/wiki/Vicugna_vicugna, http://creativecommons.org/licenses/by-sa/3.0/legalcode, CC by-SA 3.0, 124. Trilobite: Photo courtesy of Oscar Sanchez, Bolivian Fossils, 124. Trilobite: Photo courtesy of Oscar Sanchez, Bolivian Fossils, 124. Trilobite: Photo courtesy of Oscar Sanchez, Bolivian Fossils, 124. Shark:
http://marchmontpublicschoolrm106.wikispaces.com/sams+page,
http://creativecommons.org/licenses/by- sa/3.0/legalcode, CC by-SA 3.0, 126. Shark tooth:

http://ouramerica.wikispaces.com/Felix+and+Fernando+States, http://creativecommons.org/licenses/by-sa/3.0/legalcode, CC by-SA 3.0, 126. Megalodon comparison:
http://en.wikipedia.org/wiki/Megalodon#mediaviewer/File:Megalodon_scale.svg,
http://en.wikipedia.org/wiki/Wikipedia:Text_of_Creative_Commons_Attribution-ShareAlike_3.0_Unported_License, CC by-SA 3.0, 126. Megalodon teeth: Photo by Patrick Nurre, 126. Plesiosaur:
http://scientistshowtell.wikispaces.com/Mary+Anning, http://creativecommons.org/licenses/by-sa/3.0/legalcode, CC by-SA 3.0, 127. Fossil marine reptile bones: Photo by Patrick
Nurre, 127. Fossil sea urchin: Photo courtesy of Tim and Candey, earths.ancient.treasures, 127. Fossil sea urchin: Photo by Patrick Nurre, 127. Fossil sea urchin: Photo by Patrick Nurre, 127. Fossil barnacle: Photo by Patrick Nurre, 128. Barnacles: Photo by Michael Maggs.
http://en.wikipedia.org/wiki/Barnacle#mediaviewer/File:Chthamalus_stellatus.jpg,
http://creativecommons.org/licenses/by-sa/3.0/legalcode, CC by-SA 3.0, 128. Fossil worm tubes: Photo by Patrick Nurre,
128. Tube worms: Photo public domain, found at
http://en.wikipedia.org/wiki/Giant_tube_worm#mediaviewer/File:Riftia_tube_worm_colony_Galapagos_2011.jpg,
128. Fossil Crinoid: Photo by Vicki Nurre, 129. Crinoid: Photo by Alexander Vasenin.
http://en.wikipedia.org/wiki/Crinoid#mediaviewer/File:Crinoid_on_the_reef_of_Batu_Moncho_Island.JPG,
http://creativecommons.org/licenses/by-sa/3.0/legalcode, CC by-SA 3.0, 129. Dugong: Photo by Julien Willem.
http://en.wikipedia.org/wiki/Dugong#mediaviewer/File:Dugong_Marsa_Alam.jpg,
http://creativecommons.org/licenses/by-sa/3.0/legalcode, CC by-SA 3.0, 129. Marine mammal dugong fossil: Photo by Patrick Nurre, 129. Fossil horn coral: Photo by Patrick Nurre, 130. Living coral:
http://biomesfifth10.wikispaces.com/Coral+Reef+Facts, http://creativecommons.org/licenses/by-sa/3.0/legalcode, CC by-SA 3.0, 130. Fossil fish vertebrae: Photo by Patrick Nurre, 130. Enchodus: Drawing by Dmitir Bogdanov.
http://en.wikipedia.org/wiki/Enchodus#mediaviewer/File:Enchodus2.jpg, http://creativecommons.org/licenses/by-sa/3.0/legalcode, CC by-SA 3.0, 130. Fossil fish: Courtesy of Tim and Candey, earths.ancient.treasures, 131. Fossil fish: Courtesy of Tim and Candey, earths.ancient.treasures, 131. Coral cross section: Source NOAA, public domain, found at http://www.seos-project.eu/modules/coralreefs/coralreefs-c01-p01.html, http://creativecommons.org/licenses/by-nc-sa/2.0/legalcode, CC by-NC-SA 2.0, 130. Coral reef: Photo by Richard Ling.
http://en.wikipedia.org/wiki/Coral_reef#mediaviewer/File:Blue_Linckia_Starfish.JPG,
http://creativecommons.org/licenses/by-sa/3.0/legalcode, CC by-SA 3.0, 131.

Section VI, Part 2
Ocean floor: http://www.radford.edu/jtso/GeologyofVirginia/Tectonics/GeologyOfVATectonics6-2a.html, 132. Sand: http://images-of-elements.com/silicon.php, http://creativecommons.org/licenses/by-sa/3.0/legalcode, CC by-SA 3.0, 134.
Quartz sand: http://allencentre.wikispaces.com/A+Grain+of+Sand, http://creativecommons.org/licenses/by-sa/2.5/legalcode, CC by-SA 2.5, 134. Turtle on black sand beach: Photo by Vicki Nurre, 134. Black sand: Photo by Vicki Nurre,
134. Green sand beach: http://www.flickr.com/photos/dannyman/7953412258/in/photostream/,
https://creativecommons.org/licenses/by-sa/2.0/legalcode, CC by-SA 2.0, 135. Green sand: Photo by Patrick Nurre, 135. Magnetic sand: Photo by Patrick Nurre, 135.

Patrick Nurre has been a rock hound since childhood and has an extensive rock, mineral and fossil collection, having collected from all over the United States. In 2005, he started Northwest Treasures, which is devoted to designing geology kits for schools. He conducts numerous field trips each year in Washington State to such places as the Olympic Peninsula, Mt. Rainier, Mt. St. Helens, the Channeled Scablands, Mt. Baker and Whidbey Island. In addition, he also gives field trips to the volcano loop of Oregon and California, Mt. Hood volcanic area (Oregon), the eastern badlands of Montana and Yellowstone National Park. In 2012, he opened the Geology Learning Center in Mountlake Terrace, WA. He is a popular speaker at homeschool conventions, schools, and churches. Patrick currently co-pastors a church in the Seattle, Washington area.

If you would like to contact Patrick about speaking or field trips: northwestexpedition@msn.com
For a list of speaking topics: NorthwestRockAndFossil.com
Other books by Patrick Nurre – these are all also available with sample rock, mineral, and fossil kits at NorthwestRockAndFossil.com.

Rocks and Minerals for Little Eyes (PreK-3)
Fossils and Dinosaurs for Little Eyes (PreK-3)
Volcanoes for Little Eyes (PreK-3)
Rock Identification Made Easy (3-12)
Fossil Identification made Easy (3-12)
Bedrock Geology (high school)
Rocks and Minerals: The Stuff of the Earth (high school)
Volcanoes, Volcanic Rocks and Earthquakes (high school)
The Geology of Yellowstone – A Biblical Guide
Genesis Rock Solid – A Biblical View of Geology
Fossils Dinosaurs and Cave Men

Printed in Poland
by Amazon Fulfillment
Poland Sp. z o.o., Wrocław